INDEX

PREFACE

"The essential form of knowledge... is nothing but a representation of truth : for the truth of being and the truth of knowing are one, differing no more than the direct beam and the beam reflected."   *(Advancement of Learning)*

As our world enters into a series of decades with exciting and dramatic possibilities, so the Francis Bacon Research Trust begins a programme of Conferences, of which this is the edited transcript of the first of the continuing series, designed to provide a means for a progressive advancement in:

(a) a knowledge and understanding of the life of the Master initiate known as Sir Francis Bacon, who demonstrated in his life and after-life the whole sequence of the great initiations as an example and study for mankind.

(b) a comprehension of Francis Bacon's life work, which was to plan and commence a continuing world-wide philosophical revolution for the enlightenment of mankind, which he called "The Great Instauration."

(c) a study of the "New Method" given to mankind by Francis Bacon to help us all attain the goal of the Great Instauration - a precise and certain method by which to attain a high degree of accurate knowledge about things divine, human and natural, and hence true enlightenment, and to be able to put that wisdom-knowledge into practice as goodness and usefulness.

(d) a study of the Ancient Wisdom Knowledges and the Brotherhoods that have guarded and put into practice aspects of this Wisdom Knowledge throughout the Ages, working to a Master plan for the evolution of humanity and enlightenment of mankind.

(e) the practice, as well as the study, of all these matters.

All speculations or ideas about the truth must of necessity be put into practice in everyday life, both as a matter of responsibility and common-sense to do so, and also to test them thoroughly as to whether they are accurate concepts about what is true and good, or not. Truth "prints" goodness and usefulness, observes Francis Bacon: that is how one discovers whether it is truth or not, for deceptions and delusions abound in the human mind. Each person must discover and test truth for himself, and live it for the good of all. This is the hope of mankind, and the basis of the trust we must have in each other. When, in addition, we can learn to freely (but carefully) share our wisdom-knowledges amongst ourselves, and to work together as a fellowship of many talents, then the Great Instauration or Enlightenment of mankind and this planet will proceed far more rapidly and smoothly : "For men will begin to know their own strength when each provides his own part of the work, not every man the same."

A study and practice of the various traditions of Wisdom Knowledge - synthesised, clarified and updated by the continuing Great Instauration - is an essential foundation for the proper understanding of the scheme of the Great Instauration and teachings of Francis Bacon, hence much time will be devoted to this background knowledge. But it is to be hoped and expected that each earnest student of the Great Instauration will be making his or her own continual efforts to keep him/herself fit and healthy in mind, emotion and body, according to truth as he or she understands it at any one moment, and the particular path chosen; for only with such preparation can the higher wisdom-knowledges be properly shared, comprehended and put into action.

The Conferences are arranged wherever possible to celebrate the major festivals of each year, the keeping of which are considered to be of great importance for the welfare of mankind and the whole world. Many people are now engaged in the celebrating of festivals in a true spirit of love the world over. The festivals chosen for the work of the F.B.R.T. are the eight Solar festivals of the year - four "male" (the Solstices and Equinoxes) and four "female" (the Quarters or Fire festivals). These are the cyclic festivals related to the enlightenment of mankind and illumination of the earth, and are intimately associated with the Baconian-Rosicrucian work, Francis himself being allusively described by his contemporaries as "the brilliant Light-bearer", "the leader of the choir of Muses and of Phoebus", "the Verulamian demigod....... another Apollo". The Archangel Michael is now the more commonly used term for that which the Greeks called Apollo; namely, the great Spirit of enlightenment that "slays" the dragon-serpent of darkness with his "spear" of light.

The sequence of the festivals, beginning in November of each year, is not only connected with the annual cycle of nature, but also with the initiatory stages of man's own evolution, and a sequential study of the actual life of Francis Bacon over the course of the Conferences will assist our understanding of these things. The Lunar festivals, at the times of the full moons, are the festivals related to the personal purification and preparation of mankind, in readiness for the more general enlightenment and evolutionary step that can take place at the solar festivals. Although the Lunar festivals will not be publicly celebrated by the F.B.R.T., yet they should be understood as forming an important part of the overall work and study, in a more private way.

- Peter Dawkins

4

The Evolution of Human Consciousness

THE SOLAR AND LUNAR FESTIVALS

Today, as you know, is called All Hallows Day, tomorrow is All Souls Day, and yesterday was Hallowe'en or All Hallows Eve. It is one of our special festivals during the year. The Celts used to celebrate four of them as the Fire festivals, and another four as the Solstices and Equinoxes. This is one of the Fire Festivals. It used to be a movable feast. The Equinoxes and Solstices were fairly fixed by the position of the sun at particular times of the year. The Fire festivals moved within a "margin" of a fortnight over the years because they were made to relate to and coincide with a particular moon phase, probably the full moon, thereby linking the Solar cycle with the Lunar cycle.

In the Lunar cycle each full moon is a Lunar festival. The Lunar festivals were used to prepare each person for the Solar festivals. There is always the feminine and the masculine working together in life - the feminine being the receptive part that prepares the way for the spirit to come in, and then the masculine outgoing part which puts into action what we have received. The disciple-initiate offers the grail cup of his heart and soul to receive the light of the spirit. The cup is filled to overflowing, and then the disciple-initiate goes forth as a son of God to give that grail to the rest of life. The Lunar festivals and the Solar festivals are related to these two aspects. The lunar ones are usually kept in a more intimate way, in small groups or individually in meditation and prayer. The solar ones are really to be done with more ceremony and more outflow.

THE CANDLE

Now, I would like to start this whole day with lighting the candle. The candle is very symbolic. It is a symbol really of the whole oneness of life and of how nature evolves and transmutes itself from the physical and etheric realm, right up into the soular and spiritual realms. It also demonstrates the oneness of the three realms that man is composed of - Spirit, Soul and Body. But I will talk about that a bit later on.

PRAYER

As we light the candle, let us just say a prayer in our hearts, a prayer to God for the lighting of the flame in our own hearts, to let it shine. We will say it silently in our own hearts..... And we pray to God that this day we may be of service to Him. We pray that our minds may be opened and our hearts filled with understanding, and that we may go forth again renewed, to help others. Amen.

THE REASONS FOR FOUNDING THE FRANCIS BACON RESEARCH TRUST

In the F.B.R.T. Brochure is an outline of the aims of the Francis Bacon Research Trust and why it was set up, together with a brief sketch about Sir Francis Bacon and his Great Instauration - which was his work, his plan which he put into operation to bring enlightenment to mankind, 400 years ago. It gives just a very quick sketch. The F.B.R.T. was set up as an offspring of the work of the Francis Bacon Society. The F.B.S. has been doing research on Francis Bacon for nearly 100 years now, and one of the things that sparked off the Society's coming into being was a study of the Shakespeare plays and the recognition that there was far more behind them than just what could have been written by the man from Stratford, or anyone, however talented, without very much of an education and so on. There were many scholars and lawyers in the initial membership of the Francis Bacon Society, who went into these matters very deeply; and they began to discover clues as to the identity of the real author or authors. As they carried on their research they found that a great deal of untruth had grown up around Francis Bacon himself and his work, and they began to try to set this right. Francis in his day had to do many things in secret because it was very dangerous to do them outwardly; but that was only one of the reasons. Another was that the actual work itself involved a treasure trail. It is the way that the mystics have worked in the past when they have realised that God, in forming Creation through His Word, hid His Word within Creation.

In other words, when we are born into this world or into Creation as souls we don't suddenly see and understand truth. Truth is there, but we have to find it out and we spend lifetime after lifetime trying to discover it. It is there in Creation but we have to search for it, and always there are opportunities and clues to lead us on to find this hidden truth. Thus the whole process is like a treasure trail, a game of hide and seek. This the initiates of the past saw and recognised; and, knowing that they were created in the image of God, they therefore knew that to be sons or daughters of God they should try to imitate or copy God. And hence arose the type of work that they became involved with as they, understanding the wisdom of God, laboured to give it out to others. They realised that they could not give out everything that they had understood in one go; they would have to lead people on carefully, step by step, on a treasure trail, allowing others to have the thrill of discovering things for themselves. It is only in this way, of self discovery, that we can really come to know and understand the truth.

Francis simply followed the way of the initiates before him and, by working secretly (which means sacredly) he hid things in varying degrees of secrecy and left a treasure trail, step by step for each of us to find our ways into.

One of the main starts on this treasure trail is the question of authorship of the great literature of the Elizabethan Age. The scholar-poets put names for the authors that were not their own. Sometimes they invented names, pseudonyms - or they paid people for the use of their names, choosing people who could not possibly have written the work that their name was put to. In other words, for anyone who came along to read that work, either then or later, when trying to match up what was actually written with what was known of the life, character and accomplishments of the so-called author, the reader would realise that something was wrong, was incompatible. The supposed author would not match up with the writing. There then was the first clue, the first indication for the real searcher, the real thinker, leading onto this treasure trail. Will Shakspere of Stratford was simply one of the men carefully chosen and paid for the use of his name; and anyone who really thinks on this subject would realise that somehow that man himself could not have written the Shakespeare plays, that there is something more to it which the outer facade pretends to mask. Then the thinker would start looking beyond that facade and would find himself on the treasure trail.

The same thing applies with other books and writers of that period. When the authors invented pseudonyms they would invent pseudonyms which had a symbolic content - an actual name that meant something which would drop into people's minds who had been studying the language symbolically. The reader would realise that it was a symbolic name, and that was another way they would be led onto the treasure trail.

Now, for 400 years even this beginning has been kept a secret. This first coming into the work, to the treasure trail, has been kept a secret because it was meant to be a powerful test of men's wits. We now come to the time when we can openly suggest to people that a treasure trail exists and always has existed in all the great writings of the world, but particularly in the writings of the great English Renaissance; and so this is really why the Francis Bacon Research Trust has been set up. It is a separate entity to the Francis Bacon Society for various financial and legal reasons, but otherwise it is really part of the same impulse. This simply expands the work a little further. The work of the Trust is founded upon the work of the Francis Bacon Society, and owes its existence to this fact.

FRANCIS BACON AND THE BROTHERHOOD OF LIGHT

Francis was born at a certain time during the last Age which marked a period of harvesting and sowing seeds for the next Age that is to come. There is a cycle of seasons and initiation that occurs in every Age, just as there is in every year. Francis was a soul, a great soul, who came on the mission to begin to sow those seeds. Now seeds, as you know, are the result of a harvest of what has gone before. Everything is harvested that fruits. That which has gone before fruits, and then the fruits are harvested and the seed is taken out from the fruit and sown into the ground to form the growth for the next period, the next Age. But the seed has first of all to go into the darkness of the earth for a period of germination and strengthening and that period has been marked by the last 400 years or so.

Francis at an early age became aware of, or became initiated into the mission that his soul had undertaken, and he made the link with his high spirit - the spiritual source that he called his muse, which was identified with Pallas Athena by himself and his contemporaries. Pallas Athena is known as the Goddess or Mother of Wisdom. He gathered around him, step by step, other souls who had also incarnated to help him on his mission; until eventually he had around him a brotherhood, a fellowship, and that fellowship was linked with other brotherhoods in England and on the continent, and in fact throughout the world, that had inherited the esoteric traditions of the world. This whole widely-spread and ancient Brotherhood was to a certain extent reorganised at that time, ready for the Age of Aquarius, and the new seeds for the New Age were sown into that international fraternity of men and women initiates.

Now I should emphasize right at the start that the real esoteric Brotherhood of Light does not necessarily have an earthly organisation for it to be manifested through. The Brotherhood is a soul brotherhood. I am going to talk later about the Brethren of the Rose Cross, but let me mention now that there are societies and organisations on earth who call themselves brethren of the Rose Cross, or Rosicrucians. Now they may or may not be true Rosicrucians. A true Rosicrucian is a soul at a certain level of being or consciousness. It is a point of soul initiation, soul consciousness; and when you are a true brother of the Rose Cross you are linked in consciousness with all others at that same level of initiation, and you may or may not choose to work in an earthly organisation to bring the light and the healing to mankind. Sometimes it is necessary and important to form an organisation on earth for this reason, but at other times it is not: so when I talk about Rosicrucians I am not essentially talking about any earthly organisation, rather I am talking about this soul fellowship. It is these initiates, this fellowship of the soul, that was contacted by Francis in his lifetime, and they linked together on the inner planes. Of course they also met physically when necessary or possible, but most of their communicating was done on the inner planes.

IGNORANCE AND TRUTH

The great Adversary that we all have to battle with in this world is Ignorance. We are all souls trying to grow from ignorance of truth to a full understanding or comprehension of truth, and it is through ignorance that evil comes into the world. Ignorance and evil are very much associated with each other. Now there is a divine interpretation of evil which I may talk about later, if there is time, but we generally know evil as something which is harmful and destructive, and it is in this sense that I am talking about evil at the moment - as something hurtful and destructive.

The source of hurt and destruction was seen to be sheer ignorance. Consider, for instance, the Cathar brotherhood, the brotherhood of St. John and of the Christ. These were beautiful souls who gave healing and help and understanding to any who asked for it, quite freely. But there were those in the organised Churches who resented the fact that these beautiful souls seemed to know and to have something that they in the orthodox Church didn't have. And, as has probably happened to all of us at one time or another, this resentment boiled up - the resentment that somebody has got something that we haven't got, and they won't give it to us, and we don't realise that they can't give it to us simply because we have not yet found it in ourselves, in our souls.

When evil is very strong a self-willed power wells up in the person, and he then wants to be king, the ruler, the know-all. This was very stong in both the orthodox Church and the State during the last Age, and so the persecutions against the Cathars occurred, - and really the root cause of it was pure ignorance. If those in the Roman Catholic Church had known a little more about love and truth, and the beauty of what the soul can be, they would never have persecuted the Cathars, the Albigencies. They would never have even dreamt about it. But ignorance can blind people. Jesus said, "Father forgive them for they know not what they do." And that really sums up the whole problem in this world.

FORGIVENESS AND REDEMPTION

What the initiates are trying to learn is the law of forgiveness, the law of redemption. If you forgive somebody truly, you can help redeem him (or her) from his ignorance and bring enlightenment to his soul. And similarly, those who are dwelling in the darkness of ignorance have a veil, a blind in front of their face. That is the deep problem that they are involved in, trying to rise above ignorance and remove the veil. So all the work of the initiates was and is geared to combating this ignorance and bringing understanding and knowledge to the world. (In other words, "rending the veil" that hides the face of Truth.)

KNOWLEDGE

Now "knowledge" in initiate terms means more than just learning about things. It means knowing about and understanding something in a deep way, because you put it into action. You have experienced it, recognised it, and learnt from it, and therefore know and understand it. You know it right down deep in you. That is what the initiate meaning of knowledge is. So when I refer to knowledge as we go through the day I am talking about this total comprehension that comes when someone has grasped something of the truth, has put it into action and has seen that it works. Such a one knows a little bit of God.

So, the initiates worked to bring knowledge into the world. First they had to heal, then they could teach. The methods set up from the time of Jesus followed a definite pattern which had to be modified as the Age progressed and the various opportunities and drawbacks came in and out of the scene. I am not going to talk about this just at the moment, but I shall touch more broadly on this subject later.

THE EUROPEAN RENAISSANCE

By the time of Francis the great Renaissance of Europe, which had been set off by the initiates, welling out of the Cathar and Troubadour movement, had come to a fruition in France and England. In France it was cut off abruptly, absolutely cut dead by the Catholic Church, which itself was being manipulated by the King of Spain and those who had great power, who wanted to rule the world for what they thought were very good reasons; so he and others, in their real ignorance, thought that by ruling the world and forcibly making everybody be a good orthodox member of the Roman Catholic Church would be the true way, the work of God. They used all their power to try to bring about a world empire and world religion, according to how they saw it. They did not realise that that was not the true way.

In France there came a great flowering of the Renaissance, and it was cut dead by the power of Spain and the Roman Catholic Church. But in England, this precious island, the Renaissance carried on and was brought to a complete fruition in the English Renaissance - the Elizabethan and Jacobean era. It is those fruits of the whole European Renaissance that Francis and his fellowship gathered in; and there, at that point in time, the method that the initiates had previously used to help mankind to acquire the wisdom-knowledge was redesigned and placed onto a wider stage. Creating jewels of literature and art, they set up what is in effect the Western Mystery School.

THE OPEN UNIVERSITY MYSTERY SCHOOL

In the past Mystery schools have been physically located in isolated places - a little temple area - and certain people were carefully selected and chosen, and then trained by a teacher who was present physically, a guru on earth. The neophytes were taken through a series of experiences to stimulate their soul consciousness and therefore reach initiation. But what was done in the Elizabethan Renaissance was to create an international Mystery school like an Open University, that could eventually reach right around the world, and in which any true seeker or searcher, - by seeing the emblems in art, by reading the literature, by going to a play and experiencing what that play had to give, by hearing music that was designed in a certain way, - would be drawn into this vast Open University and find their teacher within, their own heart consciousness. The student wouldn't need somebody else necessarily to stand up, like I'm doing, and to talk - although of course talking, lecturing and so on are always useful - but they could, by their own studies, find their own teacher within themselves, the real teacher, the Christ, and they would go right into this amazing Mystery School, step by step progressing along the treasure hunt.

So what we are dealing with is the Western Mystery School, about which so many people keep asking the question, "Where is it? The East has its Mystery schools but we can't find one in the West." But it is right here! It has to be found with an open mind and heart, and somebody or something has to spark off the recognition which is the first requisite for admission. Well, many people here and there have been thus inspired during the last 400 years, but now we have come to a time when much more can be openly talked about this initiatory Quest in order to initiate more people on the labyrinthine treasure hunt. This is because humanity on this world has reached such a level of both education and experience through the opportunities and tribulations they have been having, that a very great many are open to receive this stimulation. Then they will find truth for themselves. And this is such a very great occasion because for a long time previous to this era humanity as a whole has not been ready for this stimulation. It is really something to be immensely thankful for.

INITIATION

Now initiation, which is what we are dealing with in this work, is simply an expansion of consciousness of the soul and a gradual perfecting of the nature of beauty and gentleness, or humanity - the soul being the consciousness or aggregate knowledge of an aspect of God, and an individual expression or "image" of the Light of God. Initiation is governed by a definite law. In the past the Law of God was known to be the Word of God, the two terms meaning the same thing. The Law is the Word, which is the Wisdom. Francis also referred to it as the Will of God, for the Holy Wisdom or Law of God is the Word of God which governs, of course, all life, all creation, and initiation is part of that creative process.

THE LAWS OF REDEMPTION, INITIATION AND NATURE

There is a Law of Nature, involving the Law of Karma, as many people call it. This Law governs the whole of natural evolution. Then there is a higher Law, a spiritual Law which is known as the Law of Redemption. When the Law of Redemption, the spiritual Law, is brought into operation within the natural world, and the Laws of Redemption and Nature fuse together, the result is the Law of Initiation. Just as spirit coming into nature (the body) creates the soul, so the Law of Redemption operating within the Law of Nature creates the Law of Initiation or that which governs our unfolding consciousness and beauty of nature.

THE SPIRIT, SOUL AND BODY

The great initiates of the past, who had reached a high consciousness of God, realised that the whole of Creation must have occurred because God wished to know Himself. Now, to know yourself you need to be able to see yourself in some sort of way, and then to study that "reflection", gradually learning about yourself, just like we take a mirror to study our faces and the clothes we are wearing, and so on. But before we can actually see a reflection in the mirror, there must first of all be a principle which can shine out our as yet invisible quality into that mirror. Then we get the reflection back.

The ancient sages saw God before Creation as something totally incomprehensible. They referred to this as No-Being . IT had no existence as such, IT had no Being and was no Thing, so the only way they could come to describe IT was as the Eternal Infinite Darkness. Eternity they associated with the Power of God (Omnipotence), Infinity they associated with the Love of God (Omnipresence), and the Darkness they associated with the Invisible Light or hidden Wisdom of God (Omniscience). In more human terms these are respectively the Divine Father, Mother and Son. This is the Trinity in Unity of hidden, unmanifested Principles that collectively desires to know Itself.

Thus, in the beginning that Godhead, that Trinity, had to radiate out its Principles, its Qualities; and once it had radiated out those Qualities IT had to create a mirror in which the reflection of those Qualities could be sent back to its Source. And so Creation was brought into being, and the first Creation was the manifested or Visible Light of God - a Radiance of all the Qualities, all the Principles of the Godhead. That was the Creation of the First Day - the Son of God, the Light of God - total spiritual Light radiating everywhere.

Having created this universal Radiance, the next step of this Creation was to bring into being a cosmic Mirror and the Reflection back. For this purpose the Universal Light or Radiance had to undergo a mutation or transformation in order to create a mirror, a reflective surface if you like, and to form a reflection of the Light coming back. So the Light formed a division in Itself. Part of it remained as the Radiance, the Spirit, the Light of God, but the rest formed two other parts, providing a reflecting medium and the result of that, the reflected Light.

So, what is deemed to have happened is that the First Light, that great Son of God known as the Christos or Messiah, divided into three Spheres known as the Spirit, which is the radiance of the Light, the Body, which is the reflective medium, and the Soul, which is the reflected Light itself coming back to its divine Source. This was depicted in a symbolic way by the mystics of the past as in this diagram. (See Diagram A.) The dot in the middle of the circle of radiating lines represents the invisible Source, and from that Source radiates out the Universal Light, symbolised by the radiating lines. This is the First-Born of God, the Christ Spirit or Messiah. The next step was the creation of the three Spheres, depicted in this next diagram. (See Diagram B.) In the Bible we have all this enumerated in the first Chapter of Genesis. On the first Day Light was created. On the second Day the Light divided into three Spheres - (i) the Upper Waters, which constitute the spiritual realm of the higher Heavens, the true Heaven of Radiant Light itself, or Christ Spirit (the Christos), (ii) the Firmament or Middle Waters, which is the celestial realm of the Soul, the Reflected Light, also known as the "World" or "Bride" (the Cosmos), and (iii) the Lower Waters, which constitute the natural realm of the Body - the lower aethers and elements and the physical world - also known as the Underworld. These then became inhabited by creatures created from the essences and substances of the realms to which they belonged.

Now those three realms are really one. Although they each appear individual, yet they are essentially united. The whole of the Christ Spirit, the Spirit of Truth, has to be revealed. The whole of that initial Creation has to become the Reflected Light which eventually returns home to its Source, Creation then reaching its fulfilment. The Reflected Light, the Soul Light, the Cosmos, is known as the Bride to the Spirit (which is the Christ Light). That Soul Light has to be reflected from, or evolved and transmuted from, the realm of Nature. When all Nature has transformed itself into that Soul Light, then the total marriage of the Christos with the Cosmos can take place. I use the word Cosmos to mean the Soul. Other people may use it in different ways, but I am using it to mean the Soul, and the Christos as the Spirit.

THE CANDLE - SYMBOL OF THE THREE REALMS

A candle is such a beautiful symbol of the whole process. In a lighted candle we have the three realms of Spirit, Soul and Body all manifested as it were physically, for us to learn from. First of all we have the wax of the candle. The wax needs to be pure in order to burn properly, and likewise the earthly matter of the body (and psyche) has first to become pure. When it has become sufficiently pure the spirit, enshrined in the first child-like spark of the soul, like the spark of the match, can ignite it and then the flame comes into being. The flame is the soul. It has a definite form. Study a flame and in that flame there is a radiance. Now some might say that the radiance is the same as the flame, but if you think on it and meditate on it, you will know that they are two different things or, rather, forms of light. The radiance is totally universal; it is everywhere as an unlimited, infinite form. The flame however has a definite and limited form, an individual form. But if it wasn't for that flame you would not be able to see that radiance, - and so it is with the spirit. The spiritual light is so radiant, so universal, that you could not see it and understand it without the individual form of the flame, and the flame would not exist without the wax which is transmuting into the flame. So you have depicted here those three realms of life, all related to each other and all dependant on each other.

THE NATURAL UNIVERSE, UNIVERSITY OF THE SOUL

Now it is the earth, the body, the natural universe, which was and is known as the training and testing ground - the place where the soul is nourished and evolved, making the spark into a flame, just as the wax nourishes and builds up the candle flame. The body in total includes the so-called etheric realms (the dense Aether) as well as the more earthly form of nature. This is what constitutes the great body of the universe that is the place of education for the soul. As things become learnt through successive incarnations, our knowledge and understanding and ability to execute is then tested until we reach the truth, and when that truth is arrived at we are at the point where the wax can transmute into flame - when the body can transmute into the true celestial form, and become part of the aggregate soul.

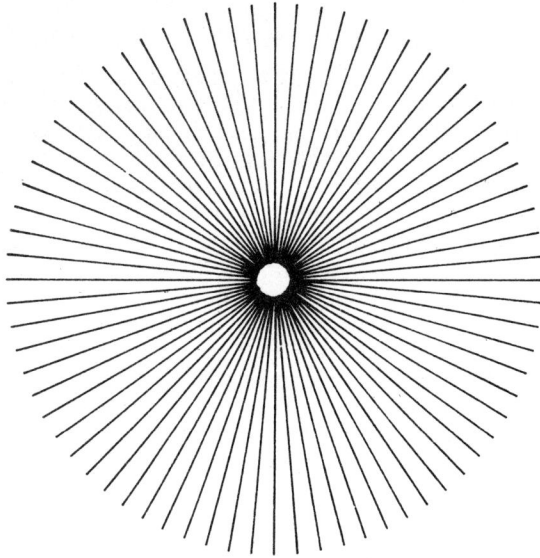

Diagram A :
Symbol of the First Light, the Christos or Messiah, the First-Born, the Radiant Manifestation of God.

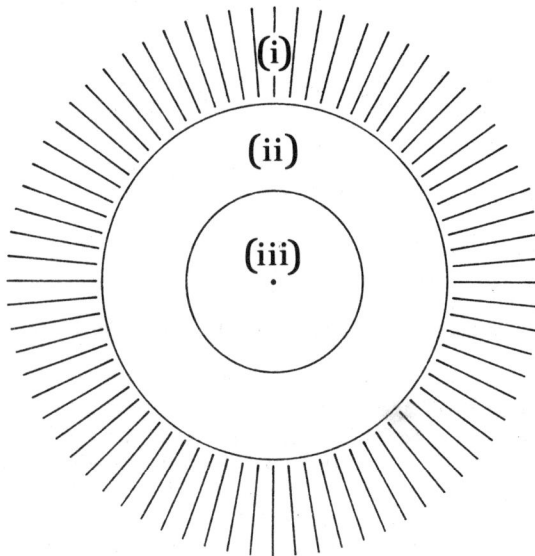

Diagram B :
Symbol of the Three Spheres of Creation

(i)   Spirit, Radiant Light, Christos, Heaven, Upper Waters, (Spiritual Aether), Spiritual (Supreme) Realm
(ii) Soul, Reflected Light, Cosmos, World or Paradise, Middle Waters or Firmament (Subtle Aether), Celestial (Superior) Realm
(iii) Body, Mirror of Nature, Universe, Underworld, Lower Waters and Earth (Dense Aether and Elements), Natural (Inferior) Realm

11

The great testing ground or field of education is set up because of two basically conflicting principles within us. You can call them by different names:- Good and Evil gives you one fundamental pair of names, or Love and Hate, or Strife and Fellowship, and so on. All the principles that exist in this evolving and transmuting universe derive from this fundamental Law of Opposites, - two conflicting principles that are striving against each other and which eventually create a union or fellowship. This was one of the basic principles taught in all the schools of initiation, the schools of the mysteries of old, and it is one of the fundamental teachings that runs through all of Francis Bacon's work, whether signed with his Baconian name or signed under another name. He loved to put together ideas which have opposites in them, or put together words which are opposite in meaning. The great poets knew this secret. They used poetry in an initiatory way by putting two words which really have different meanings side by side, and the result gives this strange power, this certain something, which enters the psyche of the reader and stimulates something exciting or profound within them. If you read the Shakespeare works, especially the Sonnets, you will come across such words and ideas which are normally in conflict with each other, and yet the poet has brought about an exciting marriage, a union of opposites. This is known as the initiatory language, the "gay science", this placing together of opposites - the language of love.

Now I am going to read you a little of what the ancients said. According to the Orphic poets:

> "Love is the great gravitating or attracting force which brought the universe into shape and gave birth to the starry spheres, out of Chaos......But its opposite power (i.e. hate or evil) is necessary to prevent everything unifying or marrying, and thus becoming One Universal Light (i.e. the Christos) until the purpose of evolution is fulfilled".

In other words we need this principle of evil or hate until the purpose of evolution is fulfilled, - and the purpose of evolution, according to the Orphic poets, is TO KNOW GOD. When all Creation knows God then the opposite principle to goodness or god-likeness, i.e. hate or evil, will disappear, and Creation will be finished.

At Eleusis also they taught, as one of the main doctrines, the separation of matter and spirit and their reconcilliation once more. The Eleusinian poets said:

> "For unity, whilst it separates from itself, identifies itself,......but when differences resolve into their source (or Union), so do they cease to exist".

So we need these opposing principles whilst we are all learning, whilst we are all evolving our souls. Francis Bacon was conveying the same secret when he wrote:

> "Strife and friendship in nature are the spurs of motion, and the Keys of Works".

Francis wrote a very great deal about strife and friendship, including a published introduction to his treatise on the subject, *The History of the Sympathy and Antipathy of Things.* Throughout all his writings we are given that same basic teaching together with the research that he had done himself. But, following the plan of the treasure trail, he does not openly talk about everything he discovered, nor everything he knows. He gives you a starter, tells you that he has completed it, but does not actually give you the whole thing. So you know that somewhere he has completed the experiment and drawn conclusions from it, and as you get to know him you will realise that it is possible to find out eventually, using one's own wit, what he both knew and accomplished, but in such a way that one knows it for oneself, having rediscovered it oneself guided by the Baconian treasure trail or labyrinth.

THE CYCLES OF NATURE AND OF INITIATION

Now let us look at some other diagrams. Diagram D is a diagram of the Cycle of Initiation. I have, for good reasons, shown it linked with the cycle of the year, the Cycle of Nature, depicted here by Diagram C. The Law of Nature brings into operation the cycles of the seasons and of the ages; and everything we do in our many lifetimes, our lifespans, operates under the same Law of Nature. Into the Law and Cycle of Nature comes the higher Law of Redemption, the spiritual law that brings enlightenment, and the result is that we obtain an initiatory cycle which links up with the seasonal events. The top section or quarter of Diagram C represents the Winter period, the Winter season of the year, and it is related to the Earth element. Next, the right-hand quarter represents the season of Spring, related to the Water element. Thirdly the bottom quarter of the diagram represents the Summer season, related to the element Air. Finally, the left-hand segment signifies the Autumn season and the element Fire. Thus we have the four seasons, Winter, Spring, Summer and Autumn, returning to Winter again, and so on and so on, in complete circles of cycle after cycle which are evolving - so they become a spiral, each cycle always coming to the same basic points as the previous ones, according to the same pattern or law, but each time at a higher level than before because something more has been learnt, and (hopefully) a little more beauty has been created.

THE SOLAR FESTIVALS

Within the cycle of four seasons are eight principal festivals - the Solar festivals - which mark the change of one season to another, and the high or mid-point of each season. In the middle of Winter, at the Winter Solstice, we have Christmas - the Festival of Birth or Rebirth. Next we come to the beginning of February and Candlemas - the Festival of Dedication - which marks the end of Winter and the start of Spring. In the middle of Spring, at the Spring Equinox (which is intimately linked with the lunar festival of Easter), we have the Festival of Promise. Then at the end of Spring comes Beltane, the Festival of Unification, coinciding both in meaning and time with the lunar Wessac Festival, and that starts off the Summer season. The mid-point of Summer is, of course, the Summer Solstice and the Midsummer Festival of Joy. At the end of Summer comes the Lammas or Festival of Transformation in August, and that begins the Autumn. At the midpoint of Autumn, the Autumnal Equinox, is the Michaelmas or Festival of Consummation. Finally we come to where we are today, at the end of Autumn and the Samhain Festival of Peace. This marks not only the beginning of Winter, but the end of one cycle or year and the start of the next.

THE CYCLE OF NATURE

In nature we see how these patterns or laws are worked out in evolving life forms. Firstly, Winter is a period when the seeds are sown in the earth, and at Christmastime they have reached a point where they can be germinated. There is an energy, a force, which goes into the seeds in the ground and brings about a quickening or germination of those seeds, and they begin to sprout underground. Then, during the last part of the winter, they slowly grow up towards the surface of the ground -but still in that earth element.

Then we come to the start of Spring, and these hidden germinated seeds in the ground suddenly begin to pop up and appear in the open air and sunlight, and Spring has truly begun. During Spring the plants grow more and more, and they start to leaf, and some of them begin to flower as well, the early ones, but generally they are going through their leafing experience.

We then come to the end of Spring, to the Beltane festival, and there is another outpouring of energy which qickens that whole plant life and it begins to flower. As I have said, some of them will have flowered before, but the general law or pattern is what this calendar is demonstrating. Nature begins to flower, and the flowering continues apace throughout the Summer.

Then we reach the Autumn, and we get the fruiting. The flower withers, but it has served its purpose:- it has created the condition that enables the fruit to grow from that flower. The flower has become fertilised and it conceives and gives birth to a "child", which is the fruit. The fruit then grows and ripens until eventually there comes the harvesting of that fruit, followed by (in the case of corn) the drying of the harvested sheaf during the last part of Autumn and then the threshing - the recovery of the seed and the burning of what is not wanted. The seed is sown back again into the ground, or else some is taken to make our bread or whatever else is needed to eat. All that is not needed and is burned is returned to the ground as humus. And so another cycle begins again, this pattern going on and on.

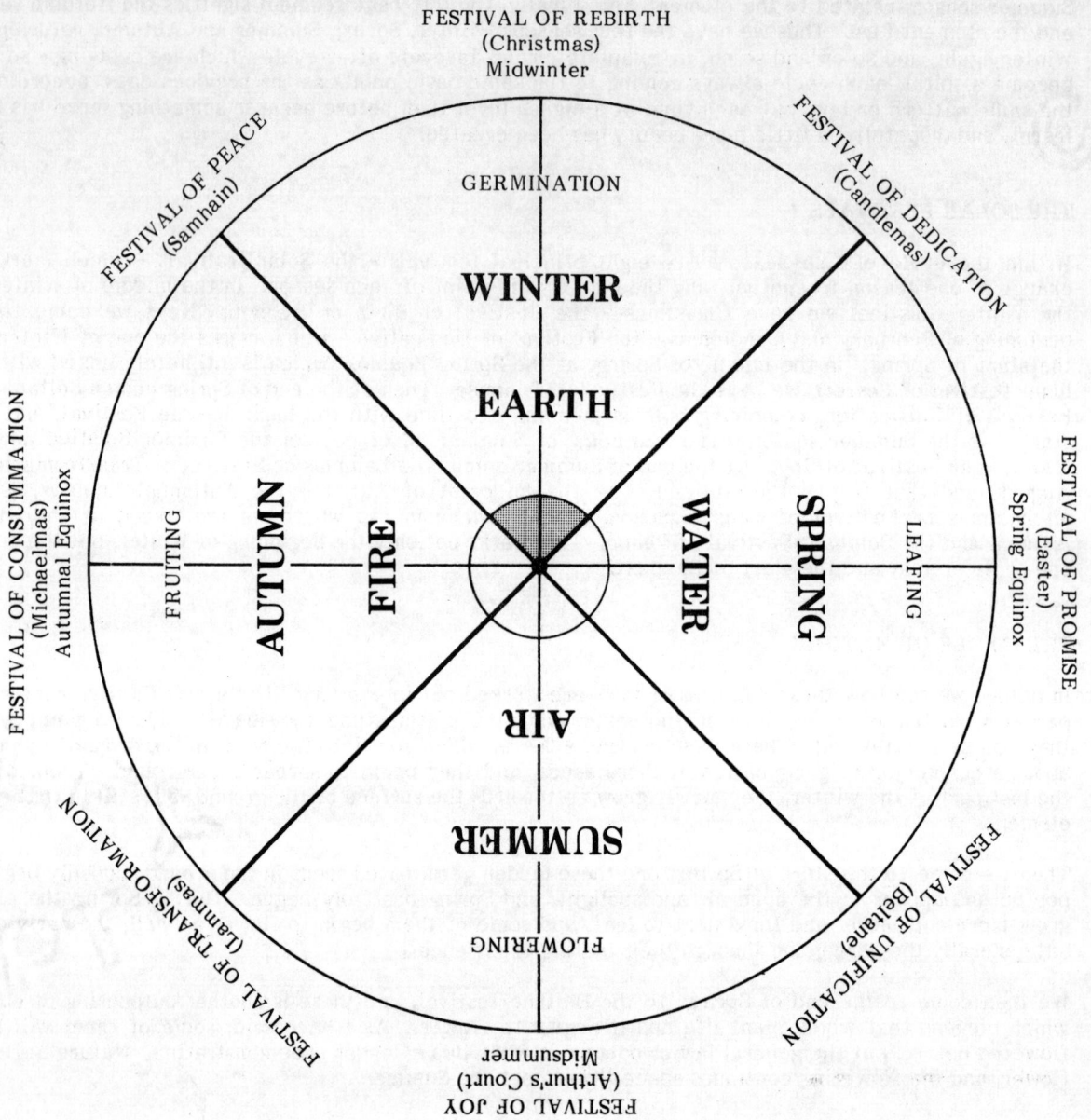

FESTIVAL OF REBIRTH
(Christmas)
Midwinter

GERMINATION

FESTIVAL OF PEACE
(Samhain)

FESTIVAL OF DEDICATION
(Candlemas)

WINTER

EARTH

FESTIVAL OF CONSUMMATION
(Michaelmas)
Autumnal Equinox

FRUITING

AUTUMN

FIRE

WATER

SPRING

LEAFING

FESTIVAL OF PROMISE
(Easter)
Spring Equinox

AIR

SUMMER

FESTIVAL OF TRANSFORMATION
(Lammas)

FLOWERING

FESTIVAL OF UNIFICATION
(Beltane)

FESTIVAL OF JOY
(Arthur's Court)
Midsummer

Diagram C :

THE CYCLE OF NATURE

14

THE CYCLES OF INITIATION

The cycles of initiation follow this same natural pattern, but in addition they have in them a fusion with the higher Law of Redemption, bringing the gifts of the Spirit and enlightenment. Thus the initiatory cycles follow basically the same pattern as the natural cycles, but sparkling with jewels of higher consciousness and comprehension.

(1) PREPARATION (EARTH)

Therefore we have the start of the Cycle of Initiation marked by the festival we are celebrating today - the Festival of Peace, sometimes called the Festival of Death or the Festival of the Dead, because the seed (the tiny spark of the pure innocent soul) is sown into the ground of an earthly body and conditions - its incarnation on earth. Then Christ mass in initiatory terms is known as the point of real birth or rebirth, for at this time the germination of the seed commences. The new-born soul first senses the light, as it were, and begins its spiritual growth. Following on from this is Candlemas - the Festival of Dedication - which is the point of emergence from earthiness into light and air.

What happens is that the soul, after incarnation at the Festival of Peace, first undergoes a natural, mundane and purely earthly experience, more or less unconscious of or unstirred by the spiritual life or law. This state of affairs carries on until the Christ Mass, when suddenly the little spark of soul, buried deep in the heart of that person's psyche and physical body, and lying up to this moment dormant, is stirred or stimulated into a tiny active flame of light by the Spirit. That person senses or receives a little feeling or intuition about the spiritual light, about the truth, and decides that he (or she) must try to do something about following or obeying that inner voice. Such a person, quickened by the Spirit, is never the same again as he was before. He becomes, as it were, reborn. He becomes a candidate for initiation. Thereafter, during the last part of the initiatory Winter, the psyche of this person is thinking and feeling about that spark of soul within himself, wherein that silent voice, the Voice of God, is felt or heard as conscience or intuition. Thereafter that person's inner recognition and response to it gradually increases, until at the Candle Mass, being by then duly prepared, he understands something of what he has been feeling and is ready to dedicate his life to God, to following the Light.

(2) THE WATER INITIATION

The candidate comes forth into the light of a spiritual consciousness or higher consciousness, and having come out into the light (as an entered apprentice or neophyte) his soul consciousness begins to grow and develop, whilst he first of all has to try to come to terms with his emotions, to make them peaceful. By the time the neophyte reaches the point marked by the Spring Equinox - the Festival of Promise - he has begun to really comprehend and recognise something of truth. He perceives how life goes on, he begins to hope, he sees the promise of the evolution of the soul, he receives a vision of the path and the goal. He becomes filled with a great sense of hope and wonder for what will unfold in time.

Filled with this hope and vision of what could be, the neophyte goes on the quest to achieve this hope, this vision. He tries to find out more and more truth. Most of all he searches for his counterpart, the balancing factor, something or somebody that he knows is part of himself. He goes on a quest for the Holy Grail. In fact the whole of this period of the Water Element (i.e. the Spring) is related to those degrees of knighthood that find their fullest expression on the next higher circle of initiation, in the fifth Degree of initiation - that of the master soul. The neophyte undergoes an analogous experience, but at a much lower level of consciousness and ability. As the lesser "knight" he learns how to control the "serpent" of his emotions and, trying to be valorous and chivalrous, goes on the quest for the Holy Grail.

At Beltane - the Festival of Unification - the neophyte finds something of the Holy Grail. He discovers an essence within his own heart, and his devotional nature becomes married or united in consciousness with it. What he finds within his own heart is both love and spiritual awareness or wisdom which he knows is true and divine, because he knows his heart has become pure and virgin. Then, having found the grail in his heart, that neophyte becomes a disciple or craftsman, just as the knight becomes a priest on the higher cycle. The disciple becomes a lesser "priest", because having found the grail he knows that he must both protect and give that grail in ministry to all. His heart is the grail cup, and the grail

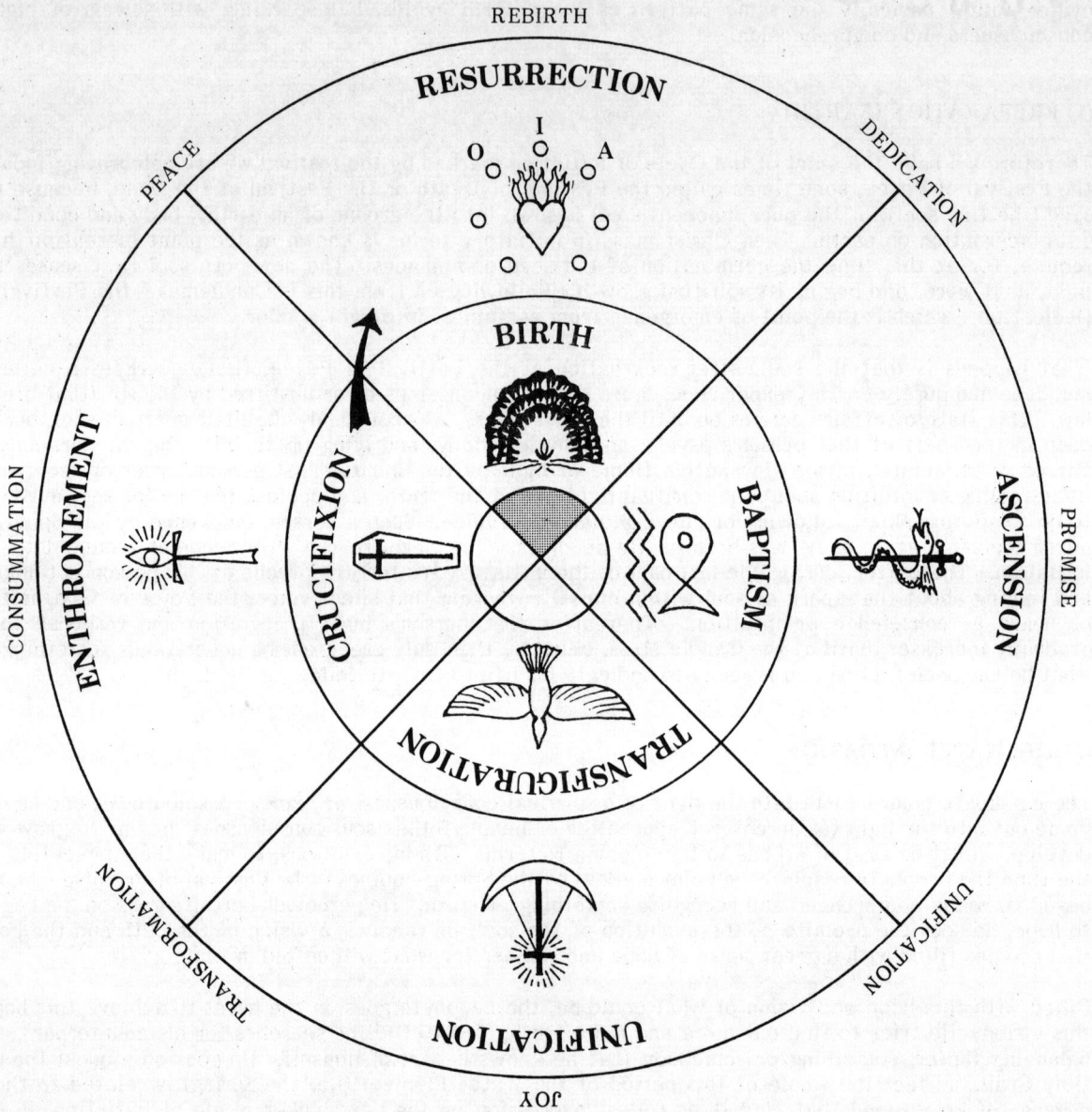

REBIRTH

RESURRECTION

O I A
O O
O O
O O

PEACE

DEDICATION

BIRTH

CONSUMMATION

ENTHRONEMENT

CRUCIFIXION

BAPTISM

ASCENSION

PROMISE

TRANSFIGURATION

TRANSFORMATION

UNIFICATION

UNIFICATION

JOY

Diagram D :

THE CYCLES OF INITIATION

is the essence of spiritual consciousness in his heart – the true soul. He knows that thenceforth he must give forth of this very essence of himself freely to others, and so Beltane marks this wonderful discovery of the Holy Grail and the beginning of a "priestly" service to mankind and all life.

(3) THE AIR INITIATION

Gradually this service builds up. The mind of the disciple becomes calmer and clearer, illumined more and more by the light from the heart, and he finds an inexpressible joy – the wonderful joy of service. A gentle energy flows through him, a light which pours through him, because of the love and wisdom in his heart and because he can give this to others and thereby help others. He sees that help actually happening, which gives a wonderful joy which the Midsummer Festival celebrates.

During the final part of the initiatory Summer the enlightening of the mind still continues, the light still pouring in and through the disciple in a glorious transfiguration, until at the end of the Summer period, at the Festival of Transformation, he reaches as full a point of transfiguration as his personality allows.

(4) THE FIRE INITIATION

At this point, the Festival of Transformation, the disciple becomes the initiate proper, the craftsman becomes the master mason. In him now can be seen that glorious kingship that will one day be manifested free from the fetters of the lower self or personality. The flame in his heart has now reached his head, surrounding it and bringing into manifestation the crown of glory – the halo or corona of Christ light, which shines about and through the face as the countenance of the Lord. As the Autumn period proceeds the initiate begins to climb the symbolic hill of transformation, the golden fire of the Christ light increasing and affecting his outer psychic and physical vehicles of incarnation considerably, transmuting their matter so that the heart becomes enlarged and strengthened.

As the initiate climbs this hill of transformation, his love or passion for God and all God's creatures increases accordingly, until at the summit he enters upon the intense passion, the blazing love for all life and truth that inevitably consumes away all that is left of dross within him, including his lower vehicles of incarnation. He knows that his personality as such has to die in order to free the greater entity that has now come to birth. The symbol of Michaelmas – the Mass of Michael and All Angels or Festival of Consummation – is of the anointed king who has climbed the sacred hill. He climbs up the hill to take his throne, and his throne is the cross. The cross is the symbol of the Law of God, the Wisdom and Will of God. The king takes up his throne or cross, and stands upon the apex of the hill as the cross (or on the cross), his arms outstretched in love and sacrifice. This symbol of the king describes the corresponding initiation at the higher level or cycle – the seventh and final initiation – but in a lesser way this is the experience of the true initiate, when the whole of the first initiatory cycle begins to reach its fulfilment. This point makes the consummation, the harvesting period. The only thing that the initiate knows that there is left for him to do AS A PERSON is a total sacrifice of all that he is as a personality.

So there is the harvesting period – the severing of the plant from the earth (i.e. all earthly attachments), and the drying out of the sheaves of corn in the fields, the stem of the plant (i.e. the psyche) to wither and die but the grain (i.e. the heart and soul) to become greatly enriched and vitalised.

Then comes Samhain, the Festival of Death or Peace, and the dried corn is threshed. The initiate is attacked by what the Greeks called the Titans, the fiery forces that come in to test and to tear apart the whole being, leaving it shredded, leaving only the heart – the golden seed or kernel of true soul. The other parts are cast away and burnt in the fire, but the heart is rescued, set free to be built up into the immortal and fully evolved soul.

THE THREE CYCLES

Here the first cycle of initiation ends and the second begins. The first cycle is that of the Lesser Mysteries, the second is that of the Greater Mysteries. One year of four seasons in the Cycle of Nature corresponds to one cycle of four initiatory periods in the Cycle of Initiation. There are eight initiatory periods in all - two cycles - consisting of seven basic initiations plus the period of preparation at the beginning. In fact, overall, there are three cycles to be considered, the very first cycle being the natural cycle where the soul is unstimulated, unawakened, and is swayed here, there and everywhere, just like the rest of nature. It is governed totally by the Law of Nature. The second cycle is the one we have just dealt with (although in its explanation I condensed the first natural cycle into the first part of the Winter season). On the second cycle the soul awakens and begins to go on the search for knowledge or illumination, which is the real start of initiation and the initiatory cycles. The third cycle concerns the initiations of the Greater Mysteries, in which the fourth initiation (Earth) is a preparation for the final three initiations, and a link between the first three and the last three initiations. If you remember the symbols used in these diagrams so far, it will help you with the next diagram in relation to what I have already shown you.

Now I know that this is a lot to take in one go at the moment, but I am saying all these things in order to sow seeds. Some of you may know all this already, and even much more, but some of you may not - so this is just to sow seeds of ideas, to give a general pattern. The rest of the Conferences, as we go on throughout this year and the next year, will go into things much more gradually, step by step, so that they can be experienced more deeply, and so that you can really come to know them within yourselves.

THE LADDER OF INITIATION

Let us continue onto the next diagram (Diagram E) - the Ladder of Initiation. Here in this diagram we are shown the seven great initiations and the predecessor, the time of preparation - eight in all, giving us that cycle of nature going around twice. (N.B. What is important to remember is that there are cycles within cycles, "wheels within wheels". The pattern of the initiatory cycle (or double cycle) is constant, but it may be experienced and manifested at a great many different levels of consciousness and ability. Everyone of us goes through these cycles in every life-time, even every day, but in terms of the greater initiations they may take many ages to complete - even aeons. So, one may become an adept or master over something, in some way, in some life-time, but it does not necessarily mean that one is a full adept or master in the true soul sense, only that one is adding a little more to the greater possibility.)

At the start we have the "first birth" experience, which is no initiation as such in terms of soul consciousness. It is a natural birth, and the soul is born into a natural body and enters into a natural life on earth. Then there comes a stimulation at the initiatory Christmas-time, and what the Christ Mass represents is the birth or quickening of that Christ child, the spiritual soul, within the heart of the person. That heart is known as the virgin, the Virgin Maria or Virgin Mary, and it forms a womb in which the Christ child can be conceived and born from.

THE CHRIST CHILD - THE ROSE FLAME OF THE HEART

The Christ child is the Christ spirit or light ensouled within an individual form - the "flame" of the heart. That flame is also known as the rose or the jewel of beauty. It is called a rose because it really does look, to spiritual sight, like a rose in bloom. In the old languages, and still in Palestine, the rose is actually known, not as a particular flower as we know it in England, but as the name given to the bud of any flower. That is the rose. It is the shape of a flame. That rose (the flame) blossoms. When the flame within the heart is born from its original spark and begins to grow, it really can be seen to blossom, and it throws off beautiful petals of light and a lovely scent comes with it. As it grows the bud or flame part gets larger and larger until eventually it embraces the head; whilst all the time the petals are being thrown off.

#	Element	Stage		Greater / Lesser Mysteries
			Peace	
7	FIRE	CHRISTED ONE (Sovereign Lord) **ENTHRONEMENT** Full illumination and union with the Christ Spirit through divine love.	Consummation	GREATER MYSTERIES
			Transformation	
6	AIR	GUARDIAN (Priest) **UNIFICATION** Discovery and knowledge of the spiritual Plan, the divine Law or Truth, in order to fulfil it.	Joy	
			Unification	
5	WATER	MASTER (Knight) **ASCENSION** Increasing mastery of the secrets of life. Wielding the sword of illumination and spiritual will.	Promise	
			Dedication	
4	EARTH	ADEPT (Arch Mason) **RESURRECTION** Emergence into a higher life and consciousness. Preparation for a higher function and work.	Rebirth	
			Peace	
3	FIRE	INITIATE (Master Mason) **CRUCIFIXION** Development of latent powers of love; compassion. Complete surrender of personal life and self.	Consummation	LESSER MYSTERIES
			Transformation	
2	AIR	DISCIPLE (Craftsman) **TRANSFIGURATION** Development of purity and harmony of thought. Illumination of the mind.	Joy	
			Unification	
1	WATER	NEOPHYTE (Entered Apprentice) **BAPTISM** Development of peace and harmony in the emotions; tranquility; patience; sympathy.	Promise	
			Dedication	
0	EARTH	CANDIDATE for admission **BIRTH** Preparation. Development of control over physical appetites; determination; courage; responsibility.	Rebirth	
			Peace	

Diagram E :

THE LADDER OF INITIATION

THE CORONA

Now, the ordinary natural rose reaches a point of bloom when it no longer has a bud-like form within the middle of all its petals, and then it dies. But the rose of the heart never dies; it always has the bud-like form within it, and it goes on blossoming more and more petals, and still it retains this wonderful bud within the centre, which rises and rises until it embraces the head. At that point something special takes place between the crown and the brow chakras. When the flame is around the head the two highest chakras create a flash between them. That flashing produces the great golden radiance of light which has been painted by great painters of the past. It is the corona -the manifestation of the crown chakra. But the corona will not manifest like that until the gentle flame of love, the rose in the heart, has formed itself around the head. The lighted candle depicts this truth, and so does the Egyptian Ankh or Cross of Life. The Ankh is a form of the Rose Cross.

With many good-natured, spiritually aware people this can happen momentarily, at certain times when they experience a wonderful feeling of unification and love for all life. They will, for that moment when they are feeling that love, have the grail flame up around their heads in a blaze of light, and then it will die down again. But as the disciple progresses along the path he becomes increasingly able to maintain such a flame and radiance to a much greater extent, until eventually it is there all the time: and that represents a wonderful consciousness of the higher levels of being and illumination of the mind.

THE FIRST INITIATION - BAPTISM

So, at Christmas is born that little flame, the first-born of the heart, the rose-bud that first appears in the heart, the Christ child. The Christ child is tended very carefully in the cave of the heart by the virgin. Then comes the beginning of Spring and the Festival of Dedication. Up until then the soul has been a candidate for initiation into the mysteries of light. When it gets to the point of dedication it actually enters into an understanding of the mysteries, and initiation has begun. The soul undergoes its first stage of initiation and becomes a neophyte or entered apprentice. As a neophyte he goes through the first great initiation which is related to the Water element, and the act of initiation is called the one of Baptism. This is a real dedication to trying to live a life of truth and becoming master of the vast sea of emotions. That means trying to control them and to purify them so that he does not get upset by the opposing forces that are all the time impinging both from within himself and also from without. We each have the conflict within ourselves, but also we are affected by outside thoughts and emotions which we pick up, which we receive, because we are linked with all life. We have to learn to cope with it all in a peaceful way.

THE SECOND INITIATION - TRANSFIGURATION

When he has reached a certain degree of control over his emotions the neophyte may then go into the next degree of initiation, which is related to the Summertime and the element of Air. Having learnt something of the Truth, mainly through feeling it with the emotions, the soul then begins to comprehend it with the mind; and the neophyte becomes the disciple, the craftsman.

THE THIRD INITIATION - RENUNCIATION OR CRUCIFIXION

After this we come into the third great initiation. This is associated with the Element Fire and the Autumn season. Having gone through the learning process with the mind, we then come to a real and severe testing. The disciple becomes the initiate proper, the master mason, and during this initiatory period he undergoes a transformation within himself. The transformation involves the harvesting; that is, the initiate is buffeted and torn to pieces psychically so that only the best part remains. On the diagram I have used here the symbol of a coffin because, in the initiations given in the Mystery schools of old, the person undergoing initiation was first of all symbolically killed and put into a coffin. That was the dramatization of the initiatory event, the play that each person seeking initiation underwent in order to symbolise and actuate what the soul had to undergo in his life-time during that period of initiation. In Egypt, after the "slaying", the initiate was immediately put to work in the Hall of the Dead, and he had to actually work with the dead, with the souls of the dead and the relatives of the dead, until he had come to understand the whole process of death and renewal. The initiate had to discover the inner fires that are within matter, because this is a Fire initiation, believe it or not! Through the death experience you are actually discovering the hidden fire.

THE FOURTH INITIATION - REVELATION OR RESURRECTION

For the Fire initiation, which involves the blooming of the heart flame, the initiate has to go deep down within himself to find the fire in the flame of the heart. Having truly found it he comes into the fourth initiation, the threshold of the Greater Mysteries, the Revelation. In the Gospels it is known as the Resurrection - resurrection out of the experience of death and fiery dissolution. The soul suddenly finds his renewal - how to renew himself in a new and wider way, the way of the adept; and so he emerges from death into this wonderful initiatory life, that of the Adept or archmason. He is received into the circle of high initiates, rising from the Hall of the Dead into the Council Chamber of the Adepts and Masters. The dots on the diagram signify the seven principals or officers of the Council, the Masters that receive the initiate into their presence as an Adept of the Brotherhood. He is then given a very special name of God, a creative name of God, and taught how to use it, when to use it and where to use it.

THE FIFTH INITIATION - ASCENSION

The fifth initiation is the one known as the Ascension, and this really takes the soul right into the Christ initiations, the Greater Mysteries, the higher initiations. The final three initiations relate to the Holy Trinity. First of all the Adept has to become a Master, a real master of himself. The initiate has had to conquer and master totally his lower personality, his lower self, and now the Adept demonstrates his complete mastery of even the Serpent or Dragon fire - the aetheric current of "fire-water" that is the very essence of all matter, all nature, and of the soul itself. This is what the Michaelic symbol shows:- the sword of divine Will piercing the dragon at the throat chakra, or cutting off the dragon's head. Really this constitutes the conquering of the dragon part of ourselves, the mastery of it - not the death of that dragon in terms of its non-existence, but the death of it as an uncontrolled essence of ourselves. The will of the Master literally becomes the Will of God.

The dragon is totally mastered and raised up on that sword. Sometimes it is a spear that is shown as piercing the dragon. The dragon or serpent, by being pierced, then starts rising up that sword or spear until it is stretched out vertically, straight and true. The Gnostics showed this as a crucifix, with a serpent wound around the shaft of the cross and its head hung over the crossbar. For the final stage they showed the serpent stretched out absolutely straight and pinned to the cross. In the same way Moses held up the brazen serpent in the wilderness. He erected a cross, and the brazen serpent was this straight serpent going up the middle of the cross, crucified on that cross but through crucifixion becoming transformed, made perfect, and revealing God's Law, the Truth. That is what Jesus symbolised to us on his cross. The process begins with the Transfiguration and Crucifixion, but is completed in the fifth initiation when the soul becomes the actual wielder of the sword or spear in full consciousness. He becomes a true knight - a Master soul - wielder of the Sword of light, Shaker of the spear of God's Will. The Druids called this stage the Bardic stage. The bardic-knight is the poet and musician "par excellence", the healer-teacher, the Bard.

THE SIXTH INITIATION - UNIFICATION

The Master decides just what work he is going to do in the service of mankind (if that is his path). Then, having decided and knowing exactly what it means in terms of responsibility, he may then go on to the next, sixth initiation, called Unification. The bardic-knight becomes a true prophet and seer, a true priest and Mediator of the Christ Light, an Ovate or "Inspired One", and goes forth in the service of mankind on the particular mission that he has chosen. This initiation is also known as the stage of the Guardian (of the grail), and in the truly greater initiations the full Master soul becomes a Guardian of one of the Rays of Christ light that radiate upon and in the hearts of humanity, giving illumination to others according to the aspect he is working on.

THE SEVENTH INITIATION - ENTHRONEMENT OR TRANSITION

After this second of the Christ initiations there is the final degree of initiation - the Transition or Enthronement. The Master becomes a fully Christed soul, a fully illumined one, which is what both the Buddha and Jesus Christ were and are. Such exalted beings are also known as the Holy Ones. This is what Druid means - the "Holy One" or "Wise One".

The Holy One or Christed One has become literally the cross of light, radiating out the light, pouring out the grail to everyone, knowing that it means a total sacrifice and that at the end of that initiation the soul will have so become the cross of divine light that there will be nothing, in a sense, left of the great Master - either as a personality, (which was surrendered at the end of the third initiation) or even as a soul, because he constantly gives his all everywhere, even as the Christ Jesus does. Yet he is all, for nothing is lost, only transformed into a higher reality. And so the Christed One, the King of light, exists everywhere in the hearts of men.

In the ancient Mystery Temples, to mark the attainment of this final initiation (at whatever level it was taken on) the Illumined One was given an Ankh or Cross of Life. In this country (Britain) you were given a little cross made of oak twigs with a piece of mistletoe wound around it. In Greece you were given a fir twig with a cone on the end and a strand of ivy wound around it. They each represent the same thing. And the candle also represents the same thing. If you put two arms on the candle, outstretched, there you have the Ankh, symbol of the glorious celestial form of the arisen and illumined soul, Truth revealed.

The Eleusinian and Dionysian Mysteries

CANDLELIGHT

"How many miles to Babylon?"
"It's three score miles and ten."
"Can I get there by candlelight?"
"O yes, and back again."

"If your heels are nimble and light
You may get there by candlelight.
If your heels are nimble and light,
O yes, and back again."

So, let us light the candle, and get there by candlelight. Because to go there, to the place of Mystery, we go in the darkness of the night, in our sleep. Sleep is used as an initiatory term for going in deep meditation to the innermost part of God within us. The sleep state was a symbol for this meditative process, as also it is a means itself, to get to "Babylon". Babylon was once, at the height of its civilization, a vibrant centre for the Mysteries of Light, long before it degenerated; so it is used as a symbol of the Temple or City of Light, with its seven sacred hills or chakra points.

The candle, as I spoke of earlier (Lecture 1), is a symbol of the at-one-ment of spirit, soul and body; so we take our candle...... We go there by candlelight - that's the only way to get there. And our heels are nimble and light, because its the heels that have the wings of Mercury-Hermes on them. If we really are walking lightly, then our heels are above the ground. We walk along like a horse does, on his toes, heels above the ground, nimble and light. It is a symbol of rising up "lightly" to the heavens, and yet still remaining in touch with the earth, thus uniting all the levels of being. And when we are happy, joyful, we do tend to walk rather lightly.

All these age-old symbols really do mean something. So once again we light the candle, and as we light it let us think of reaching into those heights and then coming back again to give what we have received to others......

HERMES AND ZEUS

Now, Hermes (or Mercury as the Romans called him) is this winged figure, the one with wings on his heels. He is known as the Messenger of God, of Zeus, the Father of Light. (Roman: Jupiter or Iu-Pater, the Father of Light.) He is a messenger of the Spirit. Light is the spiritual realm, and Zeus is the God or Principle of that realm of Light. Zeus is referred to as the Father in heaven. "Our Father, who art in heaven......" He is the Spirit, the Christos, the Light that is the Father of all worlds. That is what Zeus actually is.

Hermes, as Zeus' messenger, is able to come down from that realm of Light, the realm of Spirit, into the other realms of Soul and Body. The Light itself, as Zeus the Godhead, the Christos, is transcendant in its perfect state, and its rays are like lightning, like thunderbolts when sent down into the lower regions, consuming everything. So, initially, in order not to consume everything that is less than itself, Zeus sends out his messengers into the lower regions. Each Hermes or messenger is a ray of light, its nature made gentle and human, that is able to descend into the World and Underworld - a small ray, just enough to help those in the lower regions but not too much to burn them up.

Hermes is a symbol or name of the highly evolved soul that has reached the summit of initiation, and who elects to return again to the lesser realms as a Saviour. He is a central figure in the whole of the ancient initiations that were held in Greece. These initiations concerned what was called the Eleusinian Mysteries followed by the Dionysian Mysteries. They were brought originally from Egypt under the name of the Orphic Mysteries, Egypt having received them from Atlantis.

The Greeks, with their wonderful clarity of being able to put things into words and artistic form, really made clear for the world the Egyptian initiations which are very hard to understand, unless you happened to be initiated into the language of the time that the Egyptians used. The Greeks, with their literary and dramatic talents, made these more open for the world; and thus it is a good way to study the initiations.

The Eleusinian Mysteries are called that because thay took place at Eleusis - or at any rate Eleusis was the centre of the Mysteries, although in fact the whole of sacred Greece was involved. But Eleusis was the culminating point of the Mysteries. These Eleusinian Mysteries were known as the Lesser Mysteries, involving the initiations of the first cycle (Preparatory + 1st + 2nd + 3rd, as discussed in Lecture 1). These lead into the second initiatory cycle of the Greater Mysteries of Dionysus (4th + 5th + 6th + 7th, with the 4th initation as a linking period between the two cycles).

THE ELEUSINIAN MYSTERIES

In the recorded myth of the early Eleusinian Mysteries, besides Zeus we learn of Demeter and Hades, and the Mystery itself revolves around the Goddess Demeter.

Demeter is the Goddess of the Soul realm - the divine Principle governing it - just as Zeus is that of the Spirit realm, whilst Hades is the God or Principle governing the Body realm. The three realms of Spirit, Soul and Body were known Hermetically as Heaven, the World, and the Underworld, or Supreme, Superior and Inferior realms. Kabalistically they are the Heaven of Heavens (the spiritual sphere of Creation), Eden or Paradise (the subtle and celestial sphere of Formation), and the Earth (the natural and corporeal realm of Action and Elements). So when we talk about the World we are actually meaning the Soul or Psyche. Demeter is the Goddess of the World, of the Psyche and Soul (the Soul being the most spiritual part of the Psyche). She has, in imitation of the Holy Trinity, three aspects, so we have the Triple Goddess idea - the maiden aspect, the nymph or mother aspect, and the old crone or hag aspect. For Demeter these three aspects were named respectively as Core, who is the maiden, Persephone, who is the nymph or mother, and Hecate, who is the crone or old woman. This should be borne in mind when I talk about the myth of Demeter.

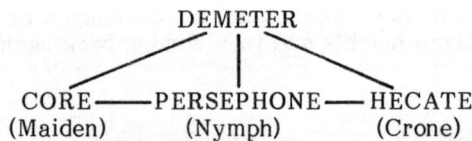

```
                          DEMETER
                             |
           _____
          |                  |                |
      CORE ———————— PERSEPHONE ———————— HECATE
     (Maiden)         (Nymph)           (Crone)
```

Demeter bore Core, the maiden, and Iacchus, the lusty one, to her brother Zeus, out of wedlock (as so often happens in the myths!). Zeus took a liking to Core, and came to her in the form of a serpent, and fertilised her. She thus became Persephone, the nymph.

At the same time Hades, looking up from the Underworld, fell in love with the beautiful Core-Persephone - a very ardent, passionate love. He did not wish to do things incorrectly or improperly, so he went to Zeus and said, in effect, "Look Zeus, can I marry Persephone? Will it be allright?"

Zeus did not want to offend Hades (because Hades was afterall his elder brother), so he did not know what to do. If he said "Yes" to Hades he would upset Demeter, his sister, but if he said "No" he would upset his brother Hades. So, very wisely, he replied that "he could neither give nor withold consent", and left it like that!

Hades took this in such a way that he thought it was a good answer, and that Zeus was tipping him the wink. He became inflamed with the idea of abducting Persephone whilst she was out gathering flowers in the field. The flowers which she was gathering were narcissi, which tells one quite a lot about the inner meanings of the story, for these flowers are the symbol of the psyche of each one of us when we are first born from the Great Mother - the psyche that is like Eve in the Garden of Eden. Persephone readily went with Hades, and descended with him into the Underworld.

Demeter was dreadfully upset at the loss of Persephone. She sought her beloved daughter for nine days and nights with a burning torch in her hand. On the tenth day, with tears in her eyes, she came to Eleusis in disguise; and there she discovered that Persephone had been abducted by Hades and was in Tartarus, the Underworld. Demeter then consulted Helios, who sees everything, and he confirmed the abduction of Persephone.

Distracted beyond measure Demeter continued to wander the world, forbidding anything to grow on the earth's surface, until eventually the whole race of men stood in danger of extinction. She besought Zeus to do something about the situation, and Zeus was thus compelled - in order that the world should become fertile again, and that the race of men should not die out - to act. The only wise thing he could do was to send Hermes down into the Underworld to talk to Hades and persuade him to release Persephone to her mother and the World.

Hermes descended into Tartarus and spoke to Hades, who very generously said that he would encourage Persephone to return to her mother, provided she hadn't actually tasted anything of his realm - because to taste the food of the Body realm gives a connection with that realm which can never be broken. The whole of the food of the Underworld is summed up in the symbol of the pomegranate, which is the symbol of life and death, and therefore if Persephone had actually eaten anything of the pomegranate then it meant that she was in fact betrothed to Hades, and that betrothal or marriage could not be severed.

Well, Hades didn't think that Persephone had actually eaten anything of the pomegranate, so he tried to persuade Persephone to go back to her mother. She did not really want to go. She was quite liking life down in the Underworld, the material world. So Hermes then started to speak to her, and eventually what he said went home to her heart, and she began to think, "Ah, I ought to go back". He awakened that love in her for her mother and for the truer things in life, so she started to go back with Hermes to the World where her mother lived.

But unfortunately Persephone had tasted something of the Underworld. She had eaten of the pomegranate, and had eaten of the very seed of the pomegranate. Some say she tasted its seven seeds, some say three - the number is symbolic. This meant that she was irrevocably betrothed to Hades. This was told by Hecate, the crone, to Demeter and Zeus, and Zeus was placed in a really difficult situation. He did not know what to do at first, because Persephone could not be severed from Hades, and yet she had promised to return to her mother - and now she actually wanted to go back to her mother! So, in the end, a compromise was effected. Persephone was to spend part of the time with her mother in the World, and part of her time with Hades in the Underworld as Queen of Tartarus.

When restored to Demeter, Persephone began to unfold and blossom, and the child in her womb began to quicken. (Do you remember how, before she went down into the Underworld, Zeus had taken a liking to her and had impregnated her? She had been carrying Zeus' child within her for all that time.) The child grew, and after seven months of pregnancy Persephone asked the child's father, Zeus, to appear before her in all his glory. Zeus knew what this would mean and so he did not want to fulfil her request; but Persephone insisted. She adamantly pleaded that she wanted to see him, and the consequences did not matter.

So Zeus appeared to her, and Persephone was consumed by his light, his fire. This great light consumed her, burnt her up; but Zeus managed to save the seven-month old child in the womb, and for another three months (these numbers are again symbolical) he kept the child sewn up in his own thigh. After those three months the child was born from Zeus' thigh, and was known as Zagreus, the horned infant.

Now that is the myth which is really a summary of the Lesser Mysteries and their initiations, and I will try to explain later how the story relates to and summarises those initiations.

THE DIONYSIAN MYSTERIES

The Greater Mysteries of Dionysus followed on from the Eleusinian Mysteries. Dionysus is also a three-fold figure: first, as the golden child, the horned (i.e. wise) infant, Zagreus; secondly, as the golden youth, Dionysus; and thirdly, as the golden man, Hermes.

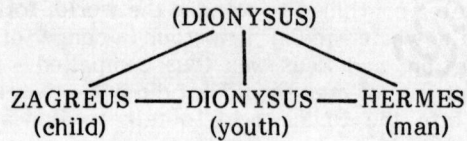

(DIONYSUS)

ZAGREUS —— DIONYSUS —— HERMES
(child)   (youth)   (man)

So, first of all the golden child is born from Zeus' thigh as the baby Zagreus. When he was born, he was delivered, they say, on a winnowing fan (or threshing floor). For, just as Persephone is likened to the corn (or to the whole of vegetation) that grows and then gives birth to the ear of corn which is the child or fruit, so Zagreus is that ear of corn (with its "horns"), and is actually born on a winnowing fan or the threshing floor, because that ear of corn is threshed and the kernels of grain separated from the husks. You may remember how, in the Gospels, John the Baptist refers to the Christ coming with the winnowing fan in his hand (Mat. 3:12, Luke 3:17), and the chaff being burnt up in unquenchable fire. It is Persephone who is consumed - she is the stalk of the plant that is dried out and burnt up in the sun (in the fields) and the husks that are burnt after the threshing.

Zagreus was so called because symbolically he was born a horned child crowned with serpents. (Think here of the symbol of Mercury, and of the Caduceus when the serpent fire is raised right up to the crown centre. The serpent writhing on the ground signifies the energies of dark matter, and of ignorance, but the serpent raised up the shaft or spear represents the energies of illumined matter, or Aether, and of wisdom.)

As soon as he was born Zagreus was attacked by the Titans sent by a jealous Hera. Hera is the spouse or feminine counterpart of Zeus, and she became awfully upset at all Zeus' doings - amours and so on - and particularly when these amours produced infants. And so Hera sent the Titans to attack Zagreus. The baby Zagreus tried to escape them by changing into the forms of a serpent, a horse, a bull, and a lion. I expect you all recognise these as the symbols of the fixed signs of the Zodiac, the horse corresponding to Aquarius. We now call Aquarius the Water Bearer, but once it had an animal symbol, and this was the horse or unicorn, emblem of the mind.

But the Titans pursued Zagreus in each of these transformations, tearing Zagreus apart in each one. (Here it is worth noting that the four fixed signs of the Zodiac form what is called the Fixed Cross. It is upon this Fixed Cross that the disciple-initiate is "fixed", and it represents his personality which he is in the act of mastering. But then, at the point marked by the birth of Zagreus, that initiate becomes the Adept, and his personality or lower self is no longer required as such. It is consumed in the fires of the heart, and the initiate passes from the Fixed Cross to the Cardinal Cross. Ordinary uninitiated humanity live upon the Mutable Cross, and are subject to the ebbs and flows and turmoils of a purely naturalistic or materialistic existence. The disciple-initiate is trying to "calm the waters", to "fix" them. The Adept uses the energies of life positively, creatively, cardinally.)

Then follow two versions of what happened next:
(a) The first story says that, having been torn to pieces, the members of Zagreus' body were placed in a cauldron and boiled, whilst a pomegranate tree sprouted from the soil where his blood had fallen. He was rescued and reconstituted by his grandmother Rhea, the Great Mother of all things, and came to life again as Dionysus.
(b) The second story relates that the Titans began to devour each part of Zagreus' dismembered body, but Pallas Athena interrupted this banquet and rescued his heart. (Note, she did not rescue the other parts, she rescued the HEART.) She enclosed this heart of Zagreus in a gypsum figure - a white figure - and breathed life into it. Zagreus was thereby reborn as Dionysus, the immortal one. All his bones, the remnants of the feast, were collected and buried at Delphi, and Zeus struck the Titans dead with a thunderbolt.

The point is that the heart, the kernel of the seed, is rescued whilst the rest is consumed and returned to the earth. Then the heart is built up into an immortal soul, an immortal being. Thus Dionysus, the immortal youth, was born.

Hermes, who was still around, temporarily transformed Dionysus into a kid or ram (i.e. Aries, one of the Cardinal signs - part of the Cardinal Cross). You know the story of the Lamb of God. The lamb is very much related to the goat, for these two symbols are complementary.

Hermes took Dionysus, the lamb of God, to the Nymphs of the Heliconian Mount Nysa, and these nymphs or mothers tended him in a cave, and fed him on honey while he grew up from childhood to youth. Whilst he was growing up and being taught by them, Dionysus discovered the secrets of the vine. He learnt how to make wine - the Holy Grail.

When he reached manhood (which he attained after discovering the Grail Secret), Hera, who up till then had been very upset about all this, actually recognised Dionysus as Zeus' son. He was then able to leave the safety of the cave and mountain, and go wandering all over the world. With Hera's recognition, he was completely free. He came out from the mountain retreat and went journeying over the world, accompanied by his tutor Silenus and a company of Satyrs and Maenads, bearing ivy-twined fir staffs tipped with pine cones (i.e. the thyrsus, or rod of initiation).

Wherever he went, Dionysus taught the secrets of the vine and wine-making. That is to say, he brought illumination to the world. It is said that he went first to Egypt, then to Mesopotamia, then to India, then back again to Europe and the Mediteranian areas. Finally he inhabited Mount Parnassus, with Apollo and Pallas Athena.

Whilst he was on these travels he met, at Naxos, the lovely Ariadne whom Theseus had deserted, and he married her.

Having established his cult throughout the world, Dionysus ascended to heaven, and now sits on the right hand of Zeus, the Father of Light, as one of the Twelve Great Ones - because Hestia (Goddess of the flame of the hearth, or heart) resigned her seat in his favour. Enthroned on the right hand of the Father of Light, Dionysus became known as Hermes, the thrice-greatest or thrice-born.

Then Hermes-Dionysus knew that there was but one choice of further action for himself to take. He knew he had to descend again to Tartarus, the Underworld, to rescue his dead mother, Persephone, and also to help begin the cycle of initiation for others. She ascended with him into Artemis' Temple, and was called Thyone. Zeus placed an apartment at her disposal.

Jesus, when he demonstrated the Christ Mysteries, acted out that final part as well, going back down into the Underworld, seeking the souls of the dead and redeeming them, releasing them and bringing them up into the Light - including the elements of his own being, his whole being. Remember that Dionysus lost quite a lot of his lower self when he was born - those parts consumed by the Titans and the bones put back into the earth. Well, all this is rescued, raised up, when Dionysus becomes Hermes and all is transmuted or transformed into light. This stage in the Christian Church has been symbolised by the Assumption of the Virgin Mary. This is one of the interpretations of the taking up of the Virgin into the heavens. It is the resurrection of the body - of the whole of natural man, even to his physical body. This is the culmination of all the Mysteries, put into mythological or symbolic story.

THE STAGES OF THE INITIATIONS

Let me show you on this diagram (Diagram F) the stages of these initiations as portrayed in the Persephone-Dionysus myth:
(1) At the bottom of the chart is the first Festival of Peace (which we are celebrating at the moment). This corresponds to the departure of Persephone for the Underworld. She is taken to the Underworld by Hades. Persephone, the psyche, is seduced by material things; yet remember, she carries within the womb of her heart the seed of Zeus, the spiritual light. The Rape of Persephone is the same as the Temptation and Fall of Eve with Adam from Paradise, descending from the celestial realm into the material realm of earthly experience and strife, in order to learn the opposites and their balance or harmony. This was the start of the Thesmophoria - the beginning of the Eleusinian Mysteries.

(2) The next stage is the Festival of Rebirth, celebrated by Christmas. Persephone actually listens to

				Peace	
7	FIRE		CHRISTED ONE (Sovereign Lord) **ENTHRONEMENT** Full illumination and union with the Christ Spirit through divine love.	Consummation	
				Transformation	
6	AIR		GUARDIAN (Priest) **UNIFICATION** Discovery and knowledge of the spiritual Plan, the divine Law or Truth, in order to fulfil it.	Joy	GREATER MYSTERIES
				Unification	
5	WATER		MASTER (Knight) **ASCENSION** Increasing mastery of the secrets of life. Wielding the sword of illumination and spiritual will.	Promise	
				Dedication	
4	EARTH		ADEPT (Arch Mason) **RESURRECTION** Emergence into a higher life and consciousness. Preparation for a higher function and work.	Rebirth	
				Peace	
3	FIRE		INITIATE (Master Mason) **CRUCIFIXION** Development of latent powers of love; compassion. Complete surrender of personal life and self.	Consummation	
				Transformation	
2	AIR		DISCIPLE (Craftsman) **TRANSFIGURATION** Development of purity and harmony of thought. Illumination of the mind.	Joy	LESSER MYSTERIES
				Unification	
1	WATER		NEOPHYTE (Entered Apprentice) **BAPTISM** Development of peace and harmony in the emotions; tranquility; patience; sympathy.	Promise	
				Dedication	
0	EARTH		CANDIDATE for admission **BIRTH** Preparation. Development of control over physical appetites; determination; courage; responsibility.	Rebirth	
				Peace	

THE LADDER OF INITIATION

(1) Dionysus descends to the Underworld as Hermes, Messenger of Light, to save Persephone. _____

(16) Dionysus is enthroned on the right hand of the Father of Light.

(15) Dionysus is taken up into heaven, to take his place in the Council of Light.

(14) Dionysus travels the world, bringing healing and illumination, teaching the secrets of the vine.

(13) Dionysus is acknowledged as a Son of Zeus (Light) by Hera.

(12) Dionysus discovers the secrets of the vine of life and how to make wine, the holy grail.

(11) Young Dionysus rededicates his life to All-Good, and learns apace.

(10) The heart of Zagreus is resurrected, and built up to become reborn as Dionysus.

(9) Persephone returns to the Underworld. Zagreus is torn to pieces by the Titans; his heart is rescued.

(8) Persephone asks to see Zeus. She is consumed with fire whilst her child, Zagreus, is born.

(7) Persephone's life begins to bear fruit. The child within her heart womb grows apace, ready for birth.

(6) Persephone becomes transfigured as light irradiates her purified mind and fills her blossoming heart.

(5) Persephone reaches a balance in her emotional life, and begins to develop her mind and open her heart.

(4) Persephone unfolds her latent gifts, and learns control over her emotions.

(3) Persephone returns to the World, and makes her vows of purity and devotion to truth.

(2) Persephone listens to Hermes and decides to return to the World. The child within her quickens.

(1) Departure of Persephone for the Underworld. _____

DIONYSIAN MYSTERIES

ELEUSINIAN MYSTERIES

Diagram F :
THE ELEUSINIAN-DIONYSIAN MYSTERIES
(i.e. THE ORPHIC MYSTERIES)

Hermes, who has been sent down in search of her. She ponders on his words. The psyche hearkens to the promptings of the voice in her heart. She decides that she will try to return to the Soul realm, and to see her mother again. That is the rebirth - the second birth. That is really the start of initiation, and it leads up through a period of preparation to:

(3) The Festival of Dedication, celebrated by Candlemas, and into the First Degree of initiation. At this point Persephone finally returns to her mother, at the beginning of Spring. In terms of natural growth and evolution, this is the reappearance of all vegetation above the ground, it having been hidden previously in the ground, in the "Underworld". At this point the Thesmophoria ended and the Lesser Eleusinia were commenced. The candidates make their vows; the psyche is dedicated to a life lived for God, or Good.

(4) Then we come to the next stage, the Festival of Promise - the Easter or Spring Festival - and the unfolding of Persephone now that she is released from Tartarus. The psyche begins to unfold its latent gifts, the vegetation its foliage. Persephone (the neophyte) begins to learn control over her passions, her emotions; and, as she completes this stage of initiation (the Baptism by Water), she becomes ready to begin the Second Degree of initiation which concerns the mind.

(5) As a balance is reached in the emotional nature, and all opposites are reconciled, "married" peacefully, so the neophyte reaches the Festival of Unification and the beginning of true discipleship. Now takes place the quickening of Persephone's mind. The mind begins to grasp the truth, and gradually this leads up to:

(6) The Festival of Joy, the Midsummer Festival when the mind of Persephone flowers - her thoughts blossom. The psyche and the heart of the disciple opens out in a wonderful expression of joy and happiness. In the natural world all vegetation flowers in a scintillating array of many colours and forms and scents. This is the Transfiguration experience, as light pours into and through the mind, opening the heart of the lotus, and it culminates in:

(7) The Festival of Transformation, the beginning of the Third Degree of initiation. All the experiences of the psyche, Persephone, as the disciple, now begin to bear fruit and ripen and she becomes the initiate. In terms of the vegetable world, the plant which has leafed and flowered now fruits and begins to ripen. The ear of corn begins to turn from green to gold.

(8) As this ripening proceeds we come to Michaelmas, the Festival of Consummation. Here the Lesser Eleusinia was consummated and the Greater Eleusinia began, because at this stage we have the harvesting of past experiences and developments. The ripened corn is harvested. Persephone asks to see Zeus in all his glory. Zeus appears to her, and consumes her, the psyche, with fire. This happens outwardly in nature, for the harvested corn is stacked in sheaves in the fields to dry out: the stems and leaves to be "consumed with fire", i.e. physically burnt by the sunlight, gaining more life. This is the Greater Eleusinia, the great initiatory experience of Renunciation or Crucifixion. The psyche of the initiate dies whilst giving birth and freedom to the true soul (Zagreus), the golden child of the heart, whilst that child is immediately kept for the further period of time in Zeus' thigh - that is, in the fertilising light of the sun. The "thigh" was the term often used for the male fertilising organ, the organ of generation; and Zeus' thigh is the sunlight which fertilises the earth, spiritually as well as physically.

(9) Following on from this is the Festival of Peace, or Samhain. At this point the whole cycle of the Lesser Mysteries of Persephone and Demeter (the Eleusinian Mysteries) come to a climax and finish. A new cycle of the Eleusinian Mysteries begins, with the start of a new Thesmophoria. Persephone has now totally died and gone down again into the earth, into the Underworld. The withered leaves and stems of the dried out corn are dug down into the ground, to return to humus, but bearing with them some of the golden grain. Thus the seeds and the humus create new life in the new cycle.

But the Greater Mysteries of Light take off from here. The corn is threshed, and the kernels separated from the husks - the wheat from the chaff. The chaff is burnt, but the grain is carefully taken to be made into bread. Zagreus, the ear of corn, is attacked by the Titans and torn (or threshed) to pieces; but the pure white essence of his heart - the kernel - is rescued (on the winnowing fan) by Pallas Athena. Hence, at this culmination of the Lesser Mysteries, the psyche or lower self of the initiate, symbolised by Persephone, is crucified. Even the outer part of the heart, the higher mind (manas) or "husk", is renounced. (Note: This is known as the Causal Body, and as the Virgin Mary.) All is totally surrendered, torn to pieces, and the lower parts are burnt up in the phoenix fires of intense passion, the all-consuming love of the heart. But the essence of the heart - the rose flame, grail of the true soul (buddhi) - is rescued and raised up immortal. The initiate becomes the true Adept.

(10) So the Greater Mysteries commence with the death of the lower self or psyche, and the birth or resurrection of the intuitional or celestial self, the soul of light. This soul is reconstituted and formed into a new, immortal being of light. The heart of Zagreus is built up by Pallas Athena, the Goddess or Divine Mother of Wisdom, to become Dionysus, the Immortal One. Christmas, the Festival of Rebirth, on this higher cycle of initiation marks the rebirth of Zagreus as Dionysus. The gentle, tender, sensitive soul of the young Adept is finally recovered from the attacks of the Titans and made strong in the "womb" of Athena, the Divine Mother. He has discovered the nature of his very soul essence, and built it up into a greater and more beautiful form of life - a pure body of light, the Golden or Christ child. The soul is resurrected or redeemed, and in the Gospels this initiation is referred to as the Resurrection; but it is also known as the Revelation, when there is "a new heaven and earth", and the soul is gradually given or learns a new, undistorted vision and knowledge of Reality, of Truth.

In this stage of initiation (the Fifth) the soul enters as an Adept into the real Christ Degrees, but just as the former "Winter" of the Lesser Mysteries was a period of preparation for the first three initiations, so this "Winter" of the higher cycle of Greater Mysteries is a corresponding period of preparation for the final three initiations. In this sense it is a mediator between the three higher and three lower Degrees of initiation - the culmination or "Arch" Degree of the Lesser Mysteries and "Gateway" to the Greater Mysteries.

(11) From the time of the Festival of Rebirth to the Festival of Dedication, Dionysus (in the myth) is being reared in the cave, growing from childhood towards manhood, learning from the nymphs and satyrs (the wise women and men, or Masters), and fed on honey, the ambrosia of the Gods (or Holy Wisdom). At the Festival of Dedication of the Greater Mysteries, the Adept rededicates his life in a far deeper sense than before, with wisdom and knowledge of what he is doing and what it means. He becomes the full Master, and begins to ascend the degrees of high or Christ consciousness - the Ascension experience. This is fully symbolised by Dionysus in his discovery of the secrets of the vine. This is total mastery. The soul discovers how to produce the holy grail from every manifestation (or vine) of life. He is known as one who is "free".

(12) This immense discovery is marked by the Festival of Promise on the higher cycle, and eventually leads the Master on to:

(13) The Festival of Unification and the Sixth Degree of initiation - the Resurrection or Unification - when the Master discovers the real unity of all things, of all life, of all Rays or Aspects of Light. At this point in the myth Dionysus is finally recognised as the Son of Zeus by Hera - that is to say, as a true Son of Light - and he may then, if he chooses, to go forth on the great Christ Mission as a Priest or Mediator to the world. Dionysus thus travels all over the world, bringing light and healing. He brings knowledge of the Holy Grail, illumination concerning the Mysteries of Light. To the world he is known as a Chohan or Lord, an Avatar or Messenger of Light. Such great souls constitute what is known as the Christ Star Circle, and they are the Watchers or Guardians of the Rays of Christ Light which pour down into the world, and into matter. There is a particular Avatar for each Age, according to which one of the seven Rays plays the principal or dominant role in that Age.

(14) In this exalted Degree of initiation there takes place a Festival of Joy, leading up to a Festival of Transformation, for there is yet a further Degree of initiation beyond even this one - the Seventh Degree of Transition or Enthronement.

(15) The Christ Mission having been fulfilled (which may perhaps take aeons of time as we know it), at the Festival of Transformation the Chohan is taken up or absorbed into Heaven as a Mahachohan or Great Lord, a Fully Christed or Illumined One, a Holy One.

(16) The Festival of Consummation, at this level of initiation, marks the enthronement of the Holy One "on the right hand of the Father Almighty (i.e. Zeus)", in the spiritual realm of Heaven, with the World as his "footstool". This Dionysus does, and then there is but one thing left for him to do as far as we are concerned, if he so chooses, and that is to descend again as Hermes to the World and Underworld, being in (or under) the World but not of it, as it is said. And so the story begins again.

(17) The Festival of Peace - the end of one story of initiation and the beginning of another - leads the Holy One through a transition onto cycles of further evolution of which we can have no idea. We can only surmise that the pattern or law is constant; that there are microcosms within macrocosms, each manifesting this same law or pattern; and what to us is the High Heaven of Spirit, of Light, may be but a "World" to a yet higher Reality.

CORRESPONDENCES

So here, briefly, is an outline of the initiations as embodied in the Eleusinian and Dionysian Mysteries. I might add that, whereas to the Thracians and Greeks the three-aspected Mother was known as Demeter and the three-aspected Son born from her was called Dionysus, to the Christians of this Age just passing the Triple Goddess or Mother has been known as Mary (hence the three Maries) and the Son has been called Jesus. In ancient Egypt the names were respectively Isis and Horus, with Osiris being a name for the Spirit or Light, which the Greeks called Zeus and the Christians referred to as "our Father in Heaven", the Christ Spirit. To the Romans these were Ceres, Bacchus and Jupiter. Thus the Lesser Mysteries in these various traditions concern the mysteries of Demeter, or Mary, or Isis, or Ceres; whilst the Greater Mysteries concern Dionysus (and Zeus), Jesus (and Christ), Horus (and Osiris), and Bacchus (and Jupiter). The name "Jesus" actually derives from the same root word ("Heru" or "Hesu") as the name "Horus". Both names mean the same - "the ensouled Light". This is the Holy Grail.

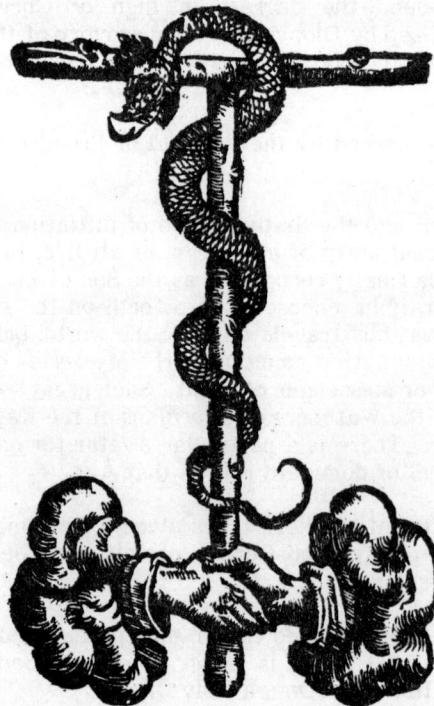

The Great Instauration of Light

THE PRAYER OF SOLOMON THE KING

"O Lord we pray above all that we might have a heart of understanding".
Amen.

THE GREAT PLAN OF THE BROTHERHOOD OF THE ROSE CROSS

Now, in this one lecture there is not very much time to talk about the Great Instauration, which is such a vast scheme. I can not possibly hope to do it any sort of justice in just this relatively short moment, so what I am doing, and what I have been trying to do all today, is to sow seeds of thought. It is a time for sowing, and so I am scattering these seeds of thought........

The Great Instauration was the name Francis Bacon gave to his life's work, that really was a work for future Ages, and in particular the Aquarian Age but also the Ages that come after that. Now this might sound a pretty fantastic claim, but Francis was - and knew he was - a high initiate of the vast Brotherhood which had been putting into operation a plan of God which they had seen, understood (as far as it is possible to do so) and then given themselves in service to bring it into effect. This Brotherhood has had many names. It was once called the Brotherhood of Pan, and Francis actually talks a little about that in his writings, quite openly; and of course there is far more not written about this ancient Brotherhood. Then it became known later as the Brotherhood of the Rose Cross. But the various names that the Brotherhood became known by publicly were derived from symbols that the Brethren had always possessed and had always worked with; so to say that the Brotherhood of the Rose Cross is a more recent phenomenon of this past Age is not entirely true. The name of the Rose Cross was made public then, but the symbol has existed and been used by that Brotherhood for a very, very long time.

THE BROTHERHOOD OF THE ATEN

What we are dealing with in the Great Instauration is a plan that was put into operation by certain of those initiates who incarnated over and over again in order to bring enlightenment to the world, and one of the first stages was to create the conditions into which a very great Ray of the Christ light might incarnate as a Soul or Son of Man and become for the world the epitomy of all the teachings, so that an explanation and synthesis of all the teachings could be seen and comprehended right in front of everyone's eyes. For a very long time this was prepared for, and in particular from the time of the 18th dynasty in Egypt which culminated with Akhenaton and his family (who were eventually murdered). The initiate Pharoahs gathered together a Brotherhood of the Rose Cross, and initiated certain chosen men and women of the Hebrew tribes that were then in Egypt. One of these Hebrews was brought up in the household of the King, given special training and recognised as a high initiate. He became the high priest of On (i.e. Heliopolis), the City of Light in Egypt, and one of the principal officers in that Brotherhood. Like some other high initiates before him he was given the initiatory name of Moses.

Also at that time there was a very great Counsellor who was the Grand Master incarnate for that Brotherhood during those critical years. His memory comes down to us in tradition as Hermes Trismegistus. He was the great guiding light behind the scenes during Akhenaton and Nefertiti's reign, and for a time afterwards. Moses became an initiate of this Brotherhood during Akhenaton's reign, and then he was given the task later on of leading out an embryonic nation that was seen to be capable, if it could fulfill its potential and karmic destiny, of becoming a true race of initiates; and hopefully, in the course of time, that race would so create the conditions on all levels, including the physical, for a very great Being to come into incarnation.

Now Moses found that he was not able to give many of these inner teachings to the motley of tribes that he had led out from Egypt, later in his life. He tried to, and the attempt was rejected outright, so symbolically he broke the tablets of the Law - which means that he reserved the teachings to be given orally or cryptically, in an esoteric way, to just the few who could understand them and make good use of them. Thus a certain select few, the chosen few who were able to understand and who were initiates of the true Is-Ra-El (or "Brotherhood of Light"), were given these precious inner teachings and interpretations, whilst the rest of the people were given outer teachings and laws to help them on their path. The outer teachings contained sufficient guide lines so that anyone who started to think more deeply would find his or her way to the real inner teachings or Caballa of knowledge. It is the ancient game of Hide and Seek once again.

THE ESSENES

The "Select" formed an inner Brotherhood within the Tribes of Israel, and later on became publicly known as the Essenes. They lived very carefully according to the Law that Moses had handed down to them - the Law that Moses himself had been given by Hermes, and which had been passed down to the Egyptians from the Atlantean Enoch. The Essenes tried to practice as pure a life as they could, linking the Angelic realm with the nature realm, and eventually they were able to create a family and a mother so pure that she could provide the conditions, free of difficult karma, into which the great Soul that we know as Jesus was able to incarnate, bringing that wonderful manifestation of the Christ even unto the physical world.

THE CHRISTINE BROTHERHOODS OF JOHN

At that marvellous time man was given everything he needed in the way of wisdom revelation to carry on his evolution. Afterwards the newly interpreted teachings were given out into the world, and certain difficulties arose - which is only natural, because of all the past karma that still had to be coped with. The real inner teachings that the Christ Jesus had given and had shown in his life could again only be shared amongst a certain few who managed to understand and practise those teachings, whilst the vast majority grasped only a small part of them. Everyone just carried on the best he could, and so there came into being the inner Christian Brotherhoods in contra-distinction to the outer, more publicly organised and controlled Churches. These inner Brotherhoods often bore the name of the Brotherhood of John, because John as a name means the Dove - the Holy Spirit or Breath of Love that bears the Word of God and brings that Word into manifestation.

THE CATHARS

These inner Christine Brotherhoods of John eventually had a period of existence in the south of France, where they became known as the Cathars, "the Pure Ones", or as the Albigencies after a province of France very much associated with them. (Note: The title of "Cathar" derives from the Greek "Katharoi", the "pure" or "holy ones", high initiates of the Orphic Mysteries, which themselves derive from the Aten Mysteries of Egypt, as do those of Israel.) The southern part of France was really a country apart from the rest of what we know as France, socially, politically and linguistically, but because of power politics within the Roman Catholic Church and within the various dynasties of Kings of northern France, these Cathars and their followers and patrons were persecuted for their lands, their influence, and for their knowledge.

The question became as to how these ancient teachings could be preserved pure and intact on earth, and how the initiates could preserve themselves and still live a life of holiness so that the truth would not be lost. The way that was chosen was to go out into the world in three different but complementary time-honoured ways, echoing the Holy Trinity once again.

THE GUILDS

One of the three ways that the initiates went out was as craftsmen, artisans. They began to help organise and inspire the great Guilds that grew up, and to introduce certain ideas and certain ceremonies that the Guilds could use which enshrined the ancient teachings about the Last Supper - the Love Feast or "Agape" - and to bring in ritual, because ritual has quite an important place in Creation. These Suppers and rituals were to create bonds of fellowship and to teach and inspire the common man to do his best manually in creating everything to be a work of art and beauty, for God and the common good.

THE TROUBADOURS

The second way was to send out artists who could teach in a way that went straight to the heart of the listener, whether they were common folk or noble gentry, and this was done through poetry and music. Hence the troubadours were sent out. They were sent out in groups of three, and the three worked together to give out the teachings and inspiration in poetry, music, acting, juggling and jesting. (The jester's part was very important.) Through the enjoyment, the pleasure that was given to others, the teachings were disseminated in a good-hearted and subtle way. The particular teachings that were chosen to be given out publicly in this way were principally the teachings of chivalry, so as to try to inspire people to chivalrous acts, to become virtuous, good mannered, kind and courageous. Once a person becomes virtuous in this way, he or she is then ready to receive the higher illuminations.

THE ORDERS OF CHIVALRY

The third way was to create a readily accessible Order or organisation that could receive these souls who had been stimulated by the troubadours or by the craft, and so the various chivalric organisations of knights grew up. The people who were stimulated by the troubadour ideas of chivalry could then join the Knights Templars or the Knights Hospitallers, and other Orders of Knights, so finding a fulfilment of their hopes and aspirations, and a channel for their energies, and being able to come gradually to the more inner teachings, the Greater Mysteries, and to be able to put them into practice through chivalry and the craft. Eventually this whole movement turned into what we know as the Renaissance, which spread across Europe and culminated in England.

1980 - 2160

That, very briefly, is a background sketch to what we are really dealing with. This plan, the Brotherhood plan, of course did not stop in there in England. It was simply put into its next stage, and is still going on now. We are now in another Age-long cycle of initiation, whose seeds of consciousness and experience (our heritage) derive from the harvest gathered in that 16th-17th Century period. We have entered into the "Winter" period of the new Age, when the seeds for the new cycle have been or are being sown into the fertile ground of humanity, or are being built up into "the Bread of life" amongst the disciples and initiates. In another 180 years from where we are now (1980) we will come to the real Christmas or Rebirth of a Golden Age, when these golden seeds which have been lying in the ground will begin to germinate, and a new world consciousness will begin to grow up in their new forms of life, to be seen in the new light of the Age. So we are now actually in a Winter period; but the early seeds are beginning to germinate, and new forms are being built up, so the work now is simply to make known a little bit more, stage by stage, of this plan and to bring more and more people to a consciousness of it so that they can play their own parts.

THE GREAT INSTAURATION

I am now going to give some quotations from Francis' writings. These are from the *Novum Organum*:

> "For we are founding in the human Intellect a true pattern of the Universe; such as it is actually found to be, not such as any one's own reason may have suggested it to him... "
> (I. Aph. 124)

"Man, the minister and interpreter of Nature, does and understands so much as he may have discerned concerning the order of Nature by observing or by meditating on facts: he knows no more, he can do no more." (I: Aph. 1)

"Neither the bare hand nor the intellect left to itself have much power; results are produced by instruments and helps; which are needed as much for the understanding as for the hand. And as instruments of use to the hand either rouse or regulate its movements, so instruments for the mind either prompt the intellect or defend it from error." (I. Aph. 2)

In this quotation Francis is showing his realisation that, just as we need certain tools in the hand to perform certain works, so the mind also needs certain types of tools to help it to understand, conceptualise and invent.

"Human knowledge and power coincide, because ignorance of the cause hinders the production of the effect. For Nature is not conquered save by obedience: and what in contemplation stands as a cause, the same in operation stands as a rule." (I. Aph. 3)

Now I have just selected these few aphorisms to give an idea of the way Francis writes (as Francis Bacon). It is not easy to understand his style of writing at first reading, (a) because it was written 400 years ago, but also (b) because he used long sentences which were very carefully constructed, and he super-imposed and put together in those sentences different ideas in a very special way.

THE NEW METHOD

But the idea of the tool for the mind inspired him to produce what he called the "New Method", and his whole work in a sense was about this new method – a new method to help people come to understand the mysteries of God. He saw and he recognised that, in Jesus Christ, mankind on this earth had been given all the wisdom that he needed for his evolution for a long, long time to come. What was lacking was an understanding or a knowledge of that wisdom. In other words people were still very much in ignorance even though the wisdom had been given, had been revealed. What was needed therefore was something to help man build up his understanding, and so the whole of the Great Instauration method is a tool to help man build up this understanding or philosophy. Francis refers to the wisdom that comes down from God into man's consciousness, via the heart, as the "Divinity" of God, and he calls the knowledge that man is building up via the use and development of his mind or intellect, and his senses, as "Philosophy."

THE PYRAMID OF PHILOSOPHY

So we have Divinity and Philosophy. Both are parts of the soul and psyche, but the one part is receiving the revelation of the light and the other part is trying to understand it and put it into operation. Thus the Great Instauration is dealing really with the philosophical aspect, the wisdom having been given through Jesus and the great prophets sufficient unto man's ends. It is Philosophy that has to be built up, and Francis saw this Philosophy as a Pyramid (or Temple) that man was trying to build up from earth to heaven. He called it **the Pyramid of the Sciences,** or **the Pyramid of Philosophy.**

In imitation of the Divine Law he made the Great Instauration to fall into six principal parts, leading to a seventh (unmentioned) part. We have talked previously about the seven major initiations which follow the same Divine Law – the seventh initiation being the culmination, the crowning of the previous six. The initiate, when he really reaches that stage, enters into the Kingdom of Light as a Son of God, enthroned at the right hand of the Father; but the six stages that preceed it are steps on the ladder of evolution leading up to that seventh stage. So first of all Bacon started his scheme in six stages which would lead to that seventh stage.

Then throughout this scheme, in all its six stages, runs a triple principle that echoes the spiritual Trinity of God which is hidden in all form. Hence the Pyramid of Philosophy can be seen as being three-aspected - a true pyramid on a triangular base. It has, standing on the ground, three sides rising up from its triangular base, expressing the Holy Trinity. These three sides deal with the three basic aspects of the Knowledge or Philosophy of man, named by Francis as **Divine Philosophy, Human Philosophy** and

Natural Philosophy. As he says:

> "In Philosophy, the contemplations of man do either penetrate unto God, – or are circumferred to Nature, – or are reflected or reverted upon himself. Out of which several enquiries there do arise three knowledges, Divine Philosophy, Natural Philosophy and Human Philosophy or Humanity. For all things are marked and stamped with this triple character of the power of God, the difference of nature and the use of man." *(Advancement of Learning)*

Now there are also two parts to Philosophy, in the sense that there is the understanding part and then the putting into action of that understanding or philosophy. In other words there is a **speculative** part and an **operative** part. The operative part is the artistic side, when we put into action the truths that we have learnt as well as we possibly can. We do it artistically because "art" means doing something well. The operative part also has three aspects to it, the counterparts to the philosophical side, and these were named by Francis as **Ecclesiastical Prudence, Human Prudence** and **Natural Prudence.**

The Pyramid of Philosophy, besides being three-faced or aspected, is also three-tiered. Francis recognised and experienced that there are three tiers or levels to which man's understanding may reach. The first tier is the realm of pure experience, the realm of the senses, the realm where we are putting thoughts and desires into action, and where we can observe whether those thoughts or desires are good or bad by their results, and by making a careful History of the experiences. The next tier is comprised of the thoughts or contemplations about those experiences and observations, and thinking about them in terms of physical laws and phenomena. This he calls **Physics** – the physics of natural, human and divine operations. The next (third) tier consists of thoughts and meditations about them on a metaphysical level, thus penetrating right up through the realm of Metaphysics unto the realm of the Godhead and the tip of the Pyramid. As for the tip or apex of the Pyramid, Francis doubted whether man would ever attain unto it – but it was worth trying all the same:

> "For Knowledges are as pyramids whereof History is the basis. So of Natural Philosophy, the basis is Natural History; the stage next the basis is Physique; the stage next the vertical point is Metaphysique. As for the vertical point, *Opus quod operateur Deus a principo usque ad finem*, the Summary Law of Nature, we know not whether man's enquiry can attain unto it. But these three be the true stages of Knowledge, and are to them that are depraved no better than the giants' hills... But to those who refer all things to the glory of God, they are as the three acclamations, *Sancte, Sancte, Sancte!* Holy in the description or dilation of His works; Holy in the connection or concantenation of them; and Holy in the union of them in a perpetual and uniform Law." *(Advancement of Learning)*

Thus we have **Divine Philosophy**, as one face of that Pyramid, being constructed of **Ecclesiastical History** as its base, **Divine Physics** as its next stage, and **Divine Metaphysics** for the third stage. **Human Philosophy**, the second facet of the Pyramid, is built up of **Civil History** (i.e. human experience), **Human Physics** and **Human Metaphysics**. **Natural Philosophy**, the third aspect of the Pyramid, has **Natural History**, **Natural Physics** and **Natural Metaphysics** for its stages. In operative terms each knowledge – Divine, Human and Natural – is put into practice in three corresponding ways: **Experimental, Mechanical** and **Magical.**

THE ADVANCEMENT OF LEARNING

Now Francis Bacon knew that he could not openly talk about all these things, except just to give brief guide lines such as the ones just mentioned. He could talk more openly about the natural and physical side of things, because men's minds were beginning to open into a study of nature; but it was still a struggle against the religions and the schoolmen of the time who tended to reject knowledge of any sort except that which was handed down either in the Church or University as dogma. A pursuit into the more magical side was then regarded as total heresy, and would have brought about a rather nasty end if discovered. The Rosicrucian brethren were very aware of the dangers, but there was more to their secrecy than that. There was the game of hide and seek. Whether the dangers had been there or not the brethren would still have followed this pattern of hiding the deeper subjects and only openly talking about the more exoteric or socially acceptable side. And so, in *The Advancement of Learning*, Francis actually paints a sketch, a plan of human knowledge as it then stood, indicating very clearly where it was deficient at that time, where it needed building up and where it was at a fairly good stage. The great thing in what Francis did then was that he gave a complete and erudite analysis of the different fields of knowledges available to man and how they relate to one another. So actually in *The Advancement of Learning* we are given a complete outline of all the areas of knowledge that man does or could attain unto and put into practice - the plan of mankind's whole philosophical work. Francis lays out, in that marvellous book, the whole plan of his field of action and thus what he is going to set into operation to try to make amends where there are deficiencies. So, when he says he is going to take all knowledge for his province and to begin to build up this pyramid in every way, he jolly well means it, even if you can't find him openly writing about it later on. Francis was always very careful about everything he said or wrote, and he always used language very, very accurately, one of the reasons being because he was largely responsible for creating the English language. (That is just a seed thought to leave you with.) Every word he used, he used because he knew exactly what it meant. He chose (or invented) each word, and put the words together in sentences very carefully to enshrine his thoughts.

Here is another quotation, just to give you an idea of what Francis said:

> "Again: some will doubt rather than object; whether we speak of perfecting by our method Natural Philosophy only, or the other sciences as well, Logic, Ethics, Politics. But we certainly understand that what we have said refers to all." (I. Aph. 127)

Here Francis is telling us that he really does intend to embrace all the sciences which he has outlined in *The Advancement of Learning*, and yet in this statement he has only gone on to refer to Logics, Ethics and Politics, which constitute Human Philosophy. You see, even in this sentence he hasn't actually gone on to mention openly Divine Philosophy, and yet he is telling us that he is referring to and developing all the aspects of the Pyramid.

I have drawn up this chart (diagram G) to help us understand what we are talking about. This two-dimensional drawing shows one face of the Pyramid of Philosophy that is built up from earth to heaven; because as Francis perceived, we cannot expect God to lower Himself, to come right down to man, with man just doing nothing about it and expecting God to come to man with no effort on man's part. Francis saw that we all had to make a good effort and to build up our understanding to reach the Godhead. This is the same teaching that the sages of old have always been saying. Francis simply restated the observation, and actually created a new method to help the mind or soul to do this. He saw, by his analysis of the state of man's knowledge, of man's understanding, that there had been quite a failure in the method of the sages of old - including methods instituted by himself in past lives, of course, because everyone reincarnates. There may be some who do not believe in reincarnation, so I apologise for mentioning this if it should offend anyone's feelings; but for me reincarnation is a reality, and for Francis it was also.

So Francis recognised his own mistakes that he had made in past lives as a sage or sages of olden times, as well as those of others, and realised that the sages had not succeeded in setting into process a method that was sufficient enough to build up man's knowledge in a really true, comprehensive and accurate way. The Greek philosophers, for instance, had set up a great movement of thought-particularly Aristotle; but the result of it was not very productive in the long run, because by the time of Bacon the universities were filled with people who discussed and conceptualised and argued endlessly about ideas they had about Aristotle's ideas, and so forth, and very little else happened. The scholars argued and argued, and yet they didn't try to put those ideas into action, they didn't try to help anybody, they didn't test those ideas out to see if they were actually true. they didn't try to improve on the ideas of Aristotle or others. They had many concepts which in fact when put to the test of experience were proved to be substantially false, whilst mankind generally was still in a state of ignorance, killing each other, torturing each other, tearing each other apart in passionate argument, ruthlessly imposing dogmas and living generally in quite a state of poverty.

This state of affairs so moved Francis that he tried to bring about, and succeeded in bringing about, a new method in the working of the Brotherhood of sages, in which truly accurate knowledge could be gained and tested by a majority, instead of a minority, of mankind.

In this chart there is depicted (as a triangle) the Pyramid of Philosophy with its three stages of History (the first stage of documented Experience), Physics (the second stage concerning material and efficient causes), and Metaphysics (the third stage concerning formal and final causes). Then there is the point of the Pyramid, indicating the pure or divine Principle of God (concerning the Supreme Cause or Summary

SUMMARY PHILOSOPHY

SPECULATIVE

DIVINE PHILOSOPHY   **HUMAN PHILOSOPHY**   **NATURAL PHILOSOPHY**

(concerning the Nature of Divinity) (concerning the Nature of Humanity) (concerning the Nature of the Universe)

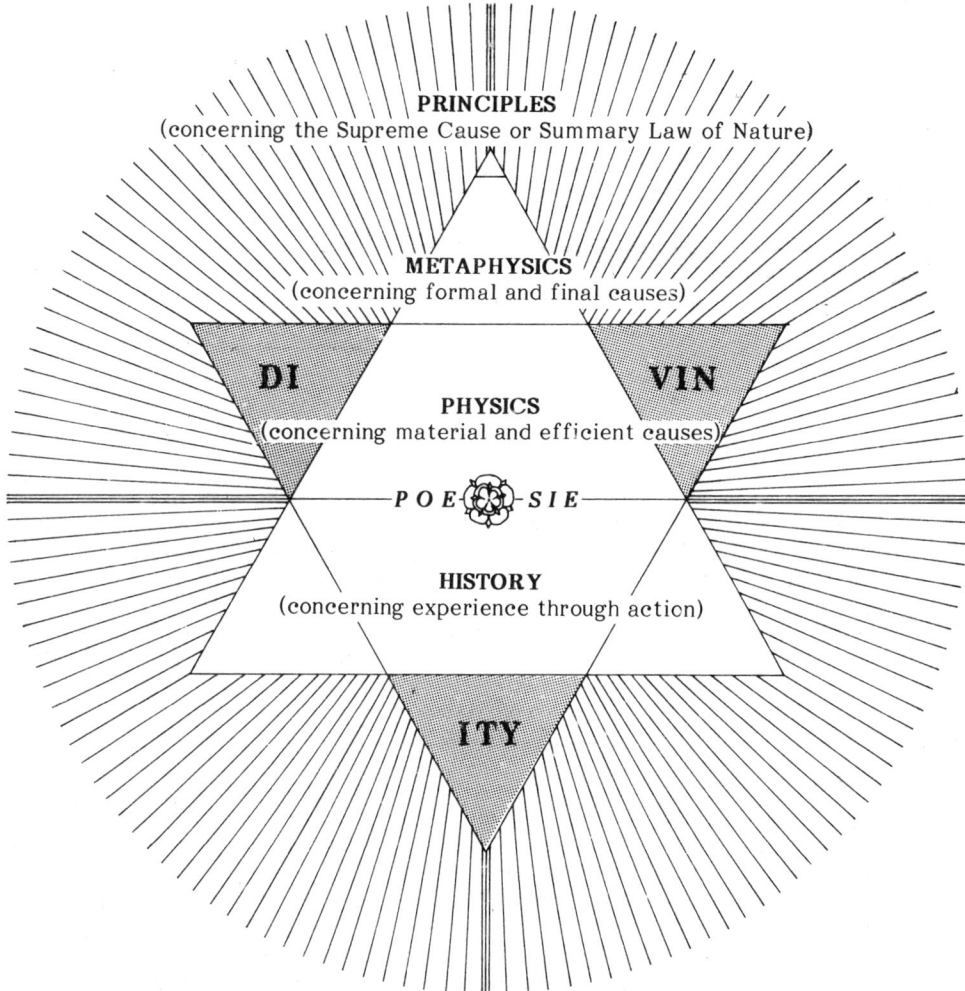

PRINCIPLES
(concerning the Supreme Cause or Summary Law of Nature)

METAPHYSICS
(concerning formal and final causes)

DI   **VIN**

PHYSICS
(concerning material and efficient causes)

P O E ⚜ *S I E*

HISTORY
(concerning experience through action)

ITY

ECCLESIASTICAL PRUDENCE   **HUMAN PRUDENCE**   **NATURAL PRUDENCE**

OPERATIVE

EXPERIMENTAL

MECHANICAL

MAGICAL

Diagram G :
THE PYRAMID OF PHILOSOPHY
(Solomon's Seal and the Christ Star)

Law of Nature). On the chart I have indicated at the top the three Speculative aspects of the Pyramid – Divine Philosophy being one side or aspect, Human Philosophy being the second side, and Natural Philosophy being the third side. At the bottom of the chart I have indicated the Operative aspects – of putting the three Speculative or Philosophical aspects into action.

THE BLAZING CHRIST STAR

Penetrating the upward pointing triangle or Pyramid of Philosophy is the down-pointing triangle or Pyramid of Divinity of the Holy Wisdom, which descends into man's soul via his heart, and illumines his mind. Here in this pattern you can recognise of course the Star of the Christ. It is traditionally known as Solomon's Seal whilst the lines are still drawn in as the two interlaced triangles. But when it blazes with light, when the mystical marriage takes place and the philosophy of man really does reflect and express the divine wisdom with which he is inspired, then those lines disappear in a fusion of light, and the Seal becomes the blazing Christ Star, the Star of Creation. The Christ Star is one of the principal symbols that the Brotherhood of Light has always worked with – the blazing Star with the Rose at the centre.

THE IMAGE OF GOD

Now, in outlining the stages of constructing this Pyramid of Knowledge, Francis has really followed what happens in the evolution of the soul. Everything has been modelled as far as humanly possible on what God has created. In everything that Francis did he tried to imitate God; and this is something that we all, ourselves, need to try and do. This is not a proud concept, but on the contrary a humble statement of reality, to know ourselves as born of God and intended to be like God in our own small ways as part of the greater whole. And so everything Francis did was modelled on what the sages of old and he himself had observed in God's Creation, both on the inner planes and in the outer worlds.

THE FIRST AND SECOND PARTS OF THE GREAT INSTAURATION

The first part of the Great Instauration was to sketch the ground plan, and most of this ground plan is given in *The Advancement and Proficience of Learning* . The second part was to actually give the method by which the Pyramid of Philosophy may be constructed, and part of this method is given in his *New Method*. Both these were translated into Latin because, Latin being a classic tongue, the words could no longer be changed in their meaning during the course of time. Thus, in order to protect the meaning of those words and thoughts, he put those works into Latin. In the *Novum Organum* (i.e. *New Method*) Francis starts his hiding. He states that he intends the *Novum Organum* to be published in three parts, but in fact only two parts were ever published. Similarly he gives us fragmented accounts of the experiments which he made for illustration of the Method. He mentions experiments which all have a bearing on light and love. He tell us in his writings that he completed these experiments, and yet when we read what was published openly we only find tantalising extracts from his experiments. But they are hints to lead us on. Enough is given to get us started on the way.

The soul follows the same pattern. When the soul prepares to come into incarnation it first of all has a consciousness of the plan, the ground plan, in which it is going to work. It is like being given a map of a country before going on a journey, in which we have somehow to get from A to B. We select our paths accordingly. The soul comes down to earth, into incarnation, with this map or plan in its heart, and has then to make its way through the environment into which it is born as best it can. Although we seem to lose knowledge of this plan, yet deep inside ourselves we retain knowledge of this map, and it is possible to gradually bring this knowledge through to our waking consciousness. In astrology we can get this ground plan interpreted for us in an astrological way.

The second stage is actually for that soul to develop a method by which it is going to get itself through that terrain in the best possible way, and so we all in our lives develop a method by which we can get through our life as well as we are capable of.

THE THIRD PART OF THE GREAT INSTAURATION

In Francis' wording the third part is concerned with "Universal Phenomena, or Natural and Experimental History with regard to an Ordered Philosophy." Having made the map or ground plan and created the method, the third part now deals with the actual first stage of building the Pyramid of Philosophy. Francis then goes on to give us illustrations or examples of how to construct this stage very accurately, using his New Method.

THE FOURTH PART OF THE GREAT INSTAURATION

The fourth part Francis refers to as "the Ladder of the Intellect" - that is, of the Understanding - called "the Thread of the Labyrinth" or "the Method of the Mind in the Comprehension of Things Exemplified", or again, "the Intellectual Sphere rectified to the Globe". This fourth part is involved with the temporary building of the second and third philosophical stages of the Pyramid upon the base of History. He wrote that "the Fourth Part of the Work is devoted to setting forth examples of Invention, in particular Subjects, choosing such Subjects as are at once the most noble and most different one from another; that there may be an example of every kind ...... I mean actual types and models by which the entire Process of the Mind and the whole Fabric and Order of Invention from the beginning to the end, in certain Subjects, and those variable and remarkable, should be set, as it were, before the Eyes."

Now, in the stage of History, Francis suggests that we construct Tables of Experience based on careful observation and experience of what actually happens in Creation, both in the outer world and in the inner realms. Following on from this, his method is not to take each experience, each observation, in isolation but to take the Tables of Experience and place the various observations into a harmonious relationship one with another, a relationship that one sees actually in Creation, and to put them into an art form that can then be acted out and presented before the eyes. In other words he intended the fourth method to be examples of experience placed dramatically in front of the eyes, echoing in condensed form inter-related aspects of Creation itself, as if seen in a mirror or frame so as to be better grasped or comprehended. This is where the works such as the Shakespeare Plays come in. They constitute Francis' example of the next stage after that of The Tables of Experience or History. All the experience that Francis and his co-helpers could gather was put together into a high dramatic and artistic form that echoes or mirrors what one can find actually in Creation. When we look at Creation it is so vast that we cannot possibly take it in very well, but as soon as we make little pictures or replicas of parts of it we can focus on that and begin to understand something of Creation.

These dramatic pictures were designed not just to appeal to the outer mind (the intellect), but also to stimulate the heart, so as to enable the heart to receive the divine wisdom or illumination. Francis is thus trying to bring into operation both parts of the Soul. He usually calls the Soul, the Mind, like the ancient philosophers did, pointing out that the Mind has an outer or rational part called the intellect, and an inner or intuitive part which is called the heart, and that the two together constitute the whole Soul or Mind.

THE FIFTH AND SIXTH PARTS OF THE GREAT INSTAURATION

So the Soul, observing these dramatic events, may then begin to understand the laws underlying them, and that understanding forms the fifth part of the Great Instauration. What the Soul can understand of the laws of truth is knowledge, and that knowledge forms the fifth part of the Instauration. But there Francis did not stop. He said in effect, it is all very well that we might obtain a wonderful drama, in which the Soul might be able to understand certain things very well, thinking that it has grasped the truth; but is it really the truth or are we still deceived? Hence we come into the sixth part, where what each Soul has understood about truth is put into action in order to test whether it really is truth; because, he says, if something is true the result must be useful, must be good, for the whole of Creation. And, of course, if something really is true it must stand the test of time, and it should continue to be good on and on and on through the years. Francis never envisaged this sixth part as being built up very quickly. He foresaw that it would take a long, long time.

The Winters Tale

Each generation is called upon to re-interpret *Shakespeare* in the light of its own world view. The cycle of plays covers the sweep of evolving consciousness of mankind. All great drama has its roots in religion. The Greek Tragedies were closely linked with the Mystery Temples. To experience them was equivalent to a catharsis of the soul, and it could be a shattering experience. The symbolism of the great myths speaks of the spiritual nature of the soul and its sufferings and transformation in the passage through earth life.

Initiation in the temple mysteries gave the soul the actual experience of its own immortal nature. One of the ordeals was the Temple Sleep, in which the candidate for initiation was laid in the tomb and the hierophant priest put him into a condition of suspended animation by withdrawing not only the astral body but most of the etheric body of vital forces. He therefore lay for three days as one dead, but soul and spirit during this period ranged widely in the spiritual worlds. Since the etheric body is the bearer of memory, the soul on its return remembered what it had experienced. It <u>knew</u> its own immortal nature. All fear of death was lifted, and it was flooded with new joy and courage when called upon to awaken. Now, this initiation experience in the Mysteries could only be given to the selected few. For the general public the teachings were given through myth and legend. The great myths enshrine the sublime truths about the soul. They speak directly to the subconscious, from the super-conscious. Hence the immense impact of the Greek drama. Hence also the power of the *Shakespeare Plays*, since on their secret hidden level they are doing just the same to us.

It is in the Comedies in particular that we find the strongest statement of esoteric truths. This is so well hidden behind the outward story that there is no need to notice it. All the plays can be interpreted on many levels, physical, psychological and spiritual. But Truth never enforces itself. It just stands, for those who care to take it, in freedom. We must learn to look at a play from the viewpoint of the myth. It speaks about soul evolution and indeed is concerned with soul experience and catharsis. Every character can be taken as a facet of the personality - yours and mine. We are the hero, our Higher Self is the heroine, the other characters represent our sub-personalities. Most people have experienced Shakesperian or Ibsonian tragedies in their own lives. They bring us the great experiences of transformation within the soul, and this is the prime purpose of life on earth.

So we are to look at *The Winters Tale* in this light. In our age the holistic world-picture is emerging strongly, bringing a conviction of the essential harmony of all life and a realization that the Universe is a vast continuum of consciousness and creative Thought. Leading scientists are now arriving at the same vision as the mystics have always held. It seems valid to take this world-view and look at a play in the light of it. Then new aspects of interpretation are given to us.

Shakespeare frequently takes an old tale and modifies it for his own purpose. The Winters Tale is a clear example. Often different levels of consciousness are indicated - Belmont and Venice, the Court and the Forest in *As You Like It*, and here in *The Winters Tale*, Sicily and Bohemia. The Kings of these two realms, Leontes and Polixenes, are shown in perfect amity. They have grown up as "puer eternas", boy eternal in closest concord. The play (like so many of the Comedies) is the story of this primal harmony shattered through human self will, to be restored in the end through true human love. Here is the essence of alchemy. This play is a mystery drama. It becomes clear that it is concerned with the Eleusinian and Dionysian mysteries.

Consider the meaning of the Persephone myth. Persophone, the soul, is carried off by Hades, Lord of the Underworld. The Underworld may be seen as the Earth realm, the plane of separation, in which souls are plunged into embodiment, gravity and the darkness of the sense world, until they can achieve understanding and redeem themselves through harmony restored. Demeter, Mother of Persephone, seeks her daughter sorrowing. She is assisted by the torch-bearer prince, Triptolemus, who risks disinheritance in order to restore her to her heavenly estate. Persephone, virgin soul, is associated with the spring of the year "when daffodils begin to peer."

The Winters Tale is the story of Demeter and Persephone, the eternal initiation of the human soul.

Shakespeare, taking the old tale from Robert Greene's "Pandosto", reverses it to suit his own purposes and makes Sicily the kingdom of Leontes, and rugged Bohemia the home of Polixenes. The secrets are hidden in the names. Hermione is the name under which both Demeter and Persephone were jointly worshipped in the Syracuse. Let us take it that Sicily is the plane of higher spiritual consciousness and that Bohemia represents the Earth level, in which all souls experience the separation and estrangement inherent in embodiment. Then we find that the name Polixenes means "Many Strangers"! What subtleties are hidden in Shakespeare's choice of names! Then why Leontes? The fourth stage of the Mithraic initiation is that of the Lion. Having passed through the trials of knowledge, silence and courage, the soul is faced with an ordeal to test its faithfulness to its own spiritual nature. This stage Leontes has reached. In the light of this, look at that astonishing and much criticised opening scene in which the King, totally inexplicably, is possessed by sudden frantic jealousy and is convinced that his dear and lifelong friend has "touched his wife forbiddenly". So violent is the jealousy that he imprisons his Queen and, when shown her new-born babe, declares that it is a bastard by Polixenes and condemns it to be abandoned and left to its fate on some barren coast far from its home. He is filled with a certainty that Polixenes has planned his murder and therefore persuades the faithful Camillo to poison him.

Look at all this in the light of an initiation test. Can Leontes remain true to his spiritual nature, represented by Hermione/Demeter? Faced with the temptation of jealousy, he fails and the whole soul is flooded with unreasoning hate, fear and fury against his wife and friend. He brings his beautiful Queen, daughter of the Emperor of Russia, to trial for her life. To his accusations she replies: "You speak a language that I understand not". He has sent to the Delphic Oracle for a ruling on the situation, but in his absolute preconception that his own judgement is right, he intends to use it simply to convince others of Hermione's guilt:

> "Though I am satisfied and need no more
> Than what I know, yet shall the oracle
> Give rest to the minds of others, such as he
> Whose ignorant credulity will not
> Come up to the truth."   (II. 1. 9)

The message from Delphi is brought during the trial, the sealed envelope opened and the document read to the Court:

> "Hermione is chaste; Polixenes blameless; Camillo a true subject;
> Leontes a jealous tyrant; his innocent babe truly begotten; and the
> King shall live without an heir till that which is lost is found."

At this, in his obsessive jealousy, Leontes defies Apollo:

> "There is no truth at all in the Oracle! The session shall proceed.
> This is mere falsehood."   (III. 2. 138)

Instantly disaster strikes. News is brought that his beloved little son Mamillius is dead as a result of the way his mother has been treated:

Leontes: "Apollo's angry and the heavens themselves
              Do strike at my injustice. Apollo, pardon
              My great profaneness 'gainst thine oracle."

But worse is to come. Paulina enters and delivers a speech of terrible and reckless abuse of the King, ending:

> "The Queen, the Queen,
> The sweetest, dearest creature's dead."

Leontes experiences complete dissolving of his jealousy and says:

> "Go on, go on:
> Thou canst not speak too much. I have deserved
> All tongues to talk their bitterest."

He declares endless repentance:

> "Once a day I'll visit
> The chapel where they lie, and tears shed there
> Shall be my recreation."

But meantime at his order Paulina's husband, Antigonus, has taken the baby girl to abandon her to her fate. We move to the stormy coast of Bohemia and there he leaves the child in her box, the "fardel", with objects to prove her royal descent and a statement that her name (given him by Hermione in a dream) is Perdita, "the lost one". Here Shakespeare gives the intriguing stage direction, "Exit, pursued by a bear". Now, no symbol in the plays is fortuitous. Antigonus means "anti-parent". The Goddess Artemis is the protectress of new-born children. She is also said at times to assume the form of a bear. So Antigonus meets his fate, the ship sinks with all hands so that all the links are destroyed. The fardel is found by a peasant and his son who take the child to their humble home.

After an interval of 16 years (which Leontes spends in repentance) we see Perdita as a grown and lovely girl in the wonderful scene of the sheep-shearing festival. Everything suggests spring. Persephone's flower is the daffodil. The rogue pedlar, Autolycus, comes in with his song:

> "When daffodils begin to peer
> With a hey the doxy over the dale
> O then comes in the spring of the year ...."

We meet the young prince Florizel, son to Polixenes, (exact contemporary with Mamillius). He has fallen in love with the beautiful daughter of the shepherd, and has dressed her up like a princess for the sheep-shearing. Shakespeare's heroines may be taken to represent the Higher Self. The object of the life experience on the earth plane is to find and unite with the spiritual aspect of our nature and so to recover the realm we have lost through the "Fall". All the Comedies, as we have said, offer variations on this theme - loss of primal harmony; fall into a plane of rivalry, conflict, separation, restriction in the sense world, the land of "many strangers"; the finding of the Higher Self through the awakening of true love; the testing ordeal to prove this love to be firm and lasting; and finally the return to the higher level of consciousness in which harmony and wholeness reigns. This is the basic theme, in a thousand variants, in the great myths, and that of the Demeter/Persephone is a fine example of it. The purpose of the Eleusinian mysteries is an initiation experience which makes conscious this alchemistic transformation of the soul.

A child in a myth represents an evolved soul-aspect of the parent. Thus Perdita, daughter to Hermione and Leontes, is their own soul development. Florizel is soul-son to Polixenes. We remember that in the complex structure of a myth all characters are in a sense aspects of the one self - yours and mine. Thus we, who have been Leontes under test, now experience the trial of Florizel.

The young prince, we have said, is like Triptolemus in his attempt to rescue Persephone. How well this is pictured in his opening words when he has dressed his princess in fitting garments:

> "These your unusual weeds to each part of you
> Does give a life; no shepherdess but Flora
> Peering in April's front. This your sheep-shearing
> Is a meeting of the petty Gods
> And you the queen on't."   (IV. 4. 1)

And she to her lover, in a speech of such beauty:

> "I would I had some flowers o' th' spring, that might
> Become your time of day - O Prosepina,
> For the flowers now that, frighted, thou let'st fall
> From Dio's wagon! Daffodils

44

That come before the swallow dares, and take
The winds of March with beauty ...." (IV. 4. 114)

Here she uses the Latin variant of the Greek name Persephone. And Florizel then describes the Higher Self and its quality:

"What you do
Still better what is done. When you speak, sweet,
I'd have you do it ever ....
                                    when you do dance I wish you
A wave o' th' sea .... Each your doing
So singular in each particular
Crowns what you are doing in the present deeds
That all your acts are queens." (IV. 4. 136)

Compare this with Bassanio's description of Portia in *Merchant of Venice*, another statement about the Higher Self:

"In Belmont is a lady richly left
And she is fair, and fairer than that word
Of wondrous virtue ....")

Now we come to Florizel's testing. Several times he is hit by the challenge of fate, each blow stonger than the last and he comes through the ordeal triumphantly. Compare this with Leontes' failure to stand up to his initiation test.

"Or, I'll be thine, my fair
Or not my father's. For I cannot be
Mine own, nor anything to any, if
I be not thine. To this I am most constant,
Though destiny say no." (IV. 4. 42)

Here speaks the self that has realised its true relation to the Self, its own spiritual principle. But his father Polixenes and the faithful old courtier, Camillo, whose place in the myth seems to be that of Conscience, appear in disguise and are welcomed as guests at the sheep-shearing. They are invited to be witnesses to the marriage of two young people and at the crisis of the scene the King reveals his identity and declares:

"Mark your divorce, young sir,
Whom son I dare not call: thou art too base
To be acknowledged .... We'll bar thee from succession
Not hold thee of our blood, no, not our kin.
Follow us to the court .... " (IV. 4. 414)

Once the King has left, Florizel, unshaken, says to the old peasant and his "daughter":

"I am sorry, not afeared,; delayed
But nothing altered: what I was I am;
        .... It cannot fail but by
The violation of my faith: Lift up thy looks.
From my succession wipe me, father,
I am heir to my affection." (IV. 4. 473)

All the plays have absolutely key lines, and here, surely, is one of them. Here indeed speaks Triptolemus, accepting disinheritance to win his love:

"Not for Bohemia, nor for the pomp that may
Be thereat gleaned, will I break my oath
To this my fair beloved." (IV. 4. 185)

We must feel that all characters, as facets of the personality, are under test. Polixenes himself is being called on to show generous sympathy for love and here he fails, as his friend Leontes had failed before him.

So Camillo, who had accompanied Polixenes in his flight from Sicily sixteen years ago, now takes it on himself to plan the return of the young couple to Leontes' Court. We move to Sicily and again meet King Leontes, ageing, chastened, continuing his daily repentance, with the loyal Paulina caring for him. News is brought of the arrival of Florizel and his princess:

> "The most peerless piece of earth, I think,
> That e'er the sun shone bright on."   (V. 1. 94)

Clearly this represents the return of Persephone from Hades to the higher plane of the Gods. Leontes greets the lovers:

> "Most dearly welcome
> Is your fair princess - Goddess. O Alas
> I lost a couple that twixt heaven and earth
> Might thus have stood, begetting wonder, as
> You, gracious couple, do .... Welcome hither
> As is the spring to th'earth."   (V. 1. 150)

But hot on their heels arrives the furious Polixenes. News of his landing and approach is brought to Leontes who asks of Florizel:

> "Is this the daughter of a King?"

>   "She is

Florizel: When once she is my wife."

Leontes: "That "once", I see by your good father's speed
> Will come on very slowly .... "

But here, in the final and most terrible trial Florizel still stands firm:

> "Dear, look up.
> Though Fortune, visible an enemy
> Should chase us, with my father, power no jot
> Hath she to change our loves."   (V. 1. 214)

He has triumphantly passed the test. So the myth can be fulfilled, for true love has come to restore the shattered harmony. We are not shown the all-too-moving scene of the revelation of the identity of Perdita. It is described by the attendant lords. Then we move to the final scene of the reuniting of Leontes with his beloved Hermione. It has been said by critics that this play is built on an impossibility and several improbabilities. The chief impossibility was Paulina's keeping the supposedly dead Queen in her house for sixteen years without the repentant King discovering. But this is a myth of Demeter and Persephone. We know that the candidates for initiation at the Eleusinian Mysteries went through a final ordeal which involved their passing through the Stygian darkness of caves from which they had to find their way. As they came through to the light, they beheld a statue of Demeter standing to receive them. And now Paulina declares that she has had an Italian sculptor carve a figure of Hermione, and she invites the group to come and admire it. The curtain is drawn and there stands the likeness of the dead Queen. Husband and daughter gaze upon it with wonder. And Paulina declares:

> "Either forbear
> Quit presently the chapel or resolve you
> For more amazement. If you can behold it,
> I'll make the statue move, indeed, descend ....
> <u>It is required</u>
> <u>You do awake your faith.</u>
> 'Tis time: descend: be stone no more.

                    Strike all that look upon with marvel. Come,
                    I'll fill your grave up.
                    Bequeath to death your numbness, for from him
                    Dear life redeems you ...."
        Leontes: "O she's warm!
                    If this be magic, let it be an art
                    Lawful as eating."

Truly this is a resurrection scene. The Oracle is fulfilled. That which was lost is found. As in so many of the Comedies, in which the alchemy of soul transformation is achieved, there are composite marriages and general rejoicing:

                    "Go together
            You precious winners all; your exultation
            Partake to everyone."

Compare this with the end of *As You Like It* with its composite marriages when Hera declares:

            "Then is there mirth in heaven
            When earthly things made even
            Atone together."

(Atonement - At-one-ment - the "integration of the personality" as spoken of by Jung. The same thought is hidden in the concept of achieving "individuality", which truly means "undividedness", reunion with the Whole.)

And Paulina, the loyal one, whose husband Antigonus had been "eaten with a bear", is given as wife to Camillo, the old figure of Conscience, which she also so well represents. So the bridge is built between rugged Bohemia, the kingdom of separation, of "many strangers", and the heavenly realm of Sicily where reign Demeter and Dionysus, in the harmony of soul relationship achieved by initiation into higher knowledge.

We need to look at that strange character Autolycus, full of roguery and song, a thief and deceiver, who says of himself:

            "Though I am not naturally honest
        I am sometimes so by chance."

He seems to represent the sub-conscious mind, like so many of Shakespeare's fools. He declares he was "littered under Mercury, who was likewise a snapper up of unconsidered trifles!" His songs and ballads and his veiled truths are the stuff of dreams, yet like the subconscious, the shadow side of our nature, he plays a major part in the drama.

And finally, there is the strange little event of the twelve Saltiers who turn up at the sheep-shearing with the offer to dance:

    Servant: "Master, there is three carters, three shepherds, three neatherds,
                three swine herds, that call themselves Saltiers and have a dance,
                a gallimaufrey of gambols."
    Polixenes: "Pray lets see these four threes of herdsmen."   (IV. 4. 322)

These in our myth may well be the Satyrs who represent the four Elements, and used to appear during the Dionysian Festival to bewail the dead god and summon him to resurrection. Shakespeare takes an old tale, itself founded on ancient myth, and reshapes it to portray the soul's initiation. Truth never constrains or enforces itself. There is never any need to enter into allegorical interpretation, for the stories have their own intrinsic delight. But to our modern minds, reaching beyond the limitations of sense-bound intellect, the attempt at spiritual interpretation of the myth may bring ideas alive so that they fire the heart with a deeper sense of the meaning in our own lives. Thus *The Winters Tale* may lead to an inner awakening of the Spring within us, a redemption of our own Persephone of the Soul. It is truly a Mystery play of Death and Resurrection, of Winter and Spring.

Symbols of the Ladder of Initiation

THE TABLE OF THE LADDER OF INITIATION (Diagram E)

The key to the table lies in the colours which relate to the Elements and in the symbols which I have drawn on the diagram. The symbols which I have used relate particularly to the Druidic Mysteries but you can find these same symbols in other countries and in their initiations. However, the emphasis of the ones I have used occur in the Druidic mysteries of Britain.

THE CAVE (EARTH): BIRTH

For instance, I have painted here a cave or grotto into which the child is first born, in the darkness and amongst the "animals" of our own lower subconscious and semi-conscious natures - the state that we are all in when we are born in our original innocence. From this birth, life becomes a quest to find light, to try and get out of that grotto of darkness and to find the light of Day (i.e. the Christ). In that grotto there is the river of nourishment, of life flowing from the great Mother of life, just as a human mother nourishes her baby. So the grotto is a symbol of the birth place. In the initiations of the past the candidate was first of all actually put into a cave or grotto beneath the ground, and there he searched his own heart for what he wanted to do, the quest for light.

FROM DARKNESS TO LIGHT

As the candidate came out of that grotto to enter into what is the first real initiation (the Baptism), beginning with the dedication service, he actually emerges "from darkness into light". These words are enshrined in the initiations given by the Orders of Freemasonry and Rosicrucianism, and others. In the First Degree of initiation, the candidate is led in blindfold from a vestibule which represents the cave. He is led into the lodge with two brethren on each side of him, who are standing trust for him. These two give their word that they have confidence in the candidate being in good faith. Then at a certain stage, and after certain preparatory stages have taken place whilst still blindfolded, the candidate is asked, "What do you most seek?" The answer the candidate gives is, "Light".

After taking an oath - that is to say, giving his word that he will not attempt to misuse or abuse the Light if given to him, and dedicating his life to serving the Light - the blindfold is removed and the candidate finds himself kneeling in front of an altar on which there is burning a light, or which bears a symbol of the Light of God. Often a bible (i.e. a holy scripture) is there on the altar, open at a meaningful page, because that represents the Light, the manifest Word or Wisdom of God. At one time in the past the whole altar stone shone with light, and sacred words and patterns were picked out in fiery outline, for the wise brethren of old understood how to create these things, and had the power then to do so. If the candidate is truly being awakened, then he will experience and perhaps see the Great Spirit or Angel of Light that overshadows the altar and lodge of brethren, and who gives the initiation.

THE MYSTERY DRAMA

Now, the process of initiation is not just an experience that happens quickly. Only a small (but vital) part occurs initially in a sudden, illuminating way - a spark to light the flame - some deep experience which stimulates and starts off the initiation. That is what these ceremonies of old did. They stimulated the initiatory process, which then continued on over quite a long period of time. The initial stimulation is given as a complete and symbolic pattern or momentary vision of what that person is then going to undergo during his period of initiation. It is framed like a drama, and in that drama each person is given "the bare bones" of everything that he is going to need to know, to guide him, after which he goes out into the world and actually lives through those experiences howsoever they may manifest for him, with the experience of the drama to help him recognise what is happening to him.

THE FISH (WATER): BAPTISM

So this is what happens at the time of dedication. The candidate goes through a drama (or "Mystery") which depicts the whole of the process through the Baptism, the first initiation. In Freemasonry, the first initiation is called the Apprentice Degree, and the dedication ceremony enables the candidate to enter the lodge as an apprentice.

The fish as symbol of the first initiation comes from the Druidic, Christian, Orphic and ancient Egyptian Mysteries. In the Celtic mythology the golden child, Taliesin, when he is first illuminated as Gwyon Bach, is actually pursued by Ceridwen, the great Mother or Dragon, because he accidentally drinks three drops from the Cauldron of wisdom and illumination which he is not intended to do. Ceridwen intends another child of hers to drink the magic potion, but whilst Gwyon is tending the Cauldron of boiling liquid three drops accidentally splash out of the Cauldron and fall onto his lips and tongue. He becomes rather scared of what Ceridwen might do, and so he gets up very quickly in order to go away and hide himself; but in doing so he knocks over the Cauldron and all the contents are spilt on the ground, leaving no chance of anyone else tasting the wonderful potion of illumination.

Gwyon races away to hide, and Ceridwen in her fury chases after him. First of all he transforms himself into a hare, signifying the start of things, the period of preparation after the birth and emergence from the grotto-like womb of Ceridwen. The hare is a wonderful and courageous creature related to the earth (and thus the Earth Element). Ceridwen chases the hare in the guise of a greyhound, and when Gwyon sees that he cannot escape her, he changes into a fish and jumps into the water. The fish lives in the water and is thus intimately related to the Water Element that has to be mastered in the first initiation. Ceridwen becomes an otter or a bigger fish that chases after him to gobble him up, so Gwyon quickly changes again into a bird and soars into the air. This is the next stage, the second initiation that is concerned with the Air Element. Ceridwen becomes a hawk and chases after him.

THE BIRD (AIR): TRANSFIGURATION

The second initiation, in Freemasonry, is known as the Craftsman's Degree. This stage is concerned with the mind and its thoughts, and is where the craftsman (once the apprentice has gone through the initial ceremony, marked by the Festival of Unification, to become passed as a craftsman) studies the seven sacred or Liberal Arts and Sciences (the amalgam or synthesis of all the arts and sciences in a sacred order and harmony) to become proficient in them. So the craftsman or disciple is enlarging and illuminating his mind, and gaining some measure of control over his thoughts.

THE TOMB (FIRE): CRUCIFIXION

The third great initiation is to do with the Fire Element. In the Celtic myth, Gwyon, chased by the hawk, metamorphoses himself from a bird to a grain of wheat; but this time he is not smart enough because Ceridwen changes herself into a hen and gobbles him up. Gwyon thus re-enters Ceridwen, the great Mother, in a state of initiatory "death", rather like going into a great tomb: and this in fact is what was done in the old initiations when the craftsman was in the process of becoming the master mason. In the Third or Master Mason Degree, the initial ceremony involves the death of the talented and proficient craftsman. He is ritually or symbolically slain and placed in a coffin or tomb, and in that tomb he undergoes an inner transmutation, in which he discovers and becomes increasingly aware of the inner fires which both enrichen his life and being, yet at the same time consume his lower self and tear him apart psychically - the result being that he finds his own true heart and soul essence. This is the Fire initiation, and it is to do with the inner passions of the heart. The Water initiation concerns the emotions, the Air is the mind, and the Fire is the heart. The heart of his being, of his true soul consciousness and power, is what the initiate discovers within himself, through the process of being torn apart psychologically, so that he can analyse himself and correct that which is found faulty or impure in the purging and transmuting fires of love.

THE HEART (EARTH-ETHER): RESURRECTION

Once the initiate discovers and masters the function and processes of his heart, he can then go on into the fourth initiation which is that of the Arch-Mason. This is really the culmination of Freemasonry, and the arch-mason is in effect the squire, the knight novitiate who is being trained to become a full knight. As such, like Theseus or any of the other heroes of old, he is given the accoutrements of knighthood to begin the quest. In the ceremony, the initiate or master-mason is raised up from the tomb, from the coffin. He is resurrected as the "heart" - the transmuted and perfected vessel of the grail of true soul consciousness, the manifest Wisdom or Word of God. This is the immortal soul, the "Golden One", called by the Celts "Taliesin". In Freemasonry this resurrection of the heart of the initiate is symbolised by the discovery and raising of the Golden Tablet or Scroll of Holy Wisdom from the vault. The full initiate or Adept knows that he is immortal in himself, the true part of himself.

THE CHAPTER

The newly-exalted arch-mason is accepted into the circle of brethren, who are symbolised in the diagram by seven dots and three letters of a sacred Word, representing the seven officers and three main principals of a chapter. A chapter is a body of Adepts constituted in a certain number and in a certain formation, of a stage, knowledge and power higher than that of a craft lodge. Whereas the first three initiations are taken in a lodge, where the numbers and formations are different, when you reach the fourth initiation you actually come into a chapter, and all knighthood and priesthood ceremonial is carried out in what is called a chapter, all within a temple. There are ten principals of a chapter, representatives of the Ten Holy Sephiroth or Principles of God. They are organised as seven officers and three grand principals. At this stage, the officers represent the seven stages of the fourth initiation, and the grand principals signify the next three initiations - knighthood, priesthood and sovereignty.

The fourth initiation is an Earth initiation. Like the previous stage of the Earth Element, this is a time of rebirth, preparation and candidature for the next three Degrees. The first rebirth is known as the "Second Birth", but this rebirth of the fourth great initiation is known as the "Third Birth", and the Adept as the "Thrice-born". The initiate, having discovered his true heart in the fires of oblation, builds up that heart into a soul that is powerful to do good deeds - in other words, an Adept. He trains himself to become a true knight, or "Phoenix". The Druids referred to this Degree as that of the Aes Dana, "the men of learning", those who have learnt and mastered as much of the basic arts and sciences of their own being and function on earth as is possible throughout the first three Degrees of initiation. In the fourth Degree the Adept puts all his talents and learning together in a synthesis, to see what he can do with them in an active and practical way. This is what the Earth initiation is all about, a putting into action all that has been learned and mastered. During this time the Adept discovers what he really knows and can do, what his particular gifts are and how they can and should be used, what his heart essence is. The discovery and integration of these things constitutes the real rebirth, the birth of Dionysus, the golden child, and from this rebirth the Adept can then proceed to the next initiation and start to put everything into full action as a true Master-soul.

ST. GEORGE AND THE DRAGON (WATER-ETHER): ASCENSION

The arch-mason becomes a knight of the grail, wielder of the sword of light. The symbol used to denote this fifth initiation is the dragon being pierced by the Michaelic sword (or spear). The dragon is a composite symbol of the four Elements, denoting the lower self or personality and the summation of the first four initiations in mastery over Water, Air, Fire and Earth. It is a symbol of the kundalini, the energies or life forces inherent in the matter of the lower self or Body nature.

ENERGY-MATTER

In matter lies the power of God. Matter was created, as Francis Bacon explains so well, "In the first act of Creation, through the power of God." In the first verse of Genesis it says, "In the beginning, God created the heavens and the earth." Now this is not the same as the heavens and the earth of life form which were created later, in the second act of Creation. What the words are trying to describe is the perfect and harmonious polarity of energy. Energy, when it polarises into a positive and negative aspect, creates matter. Matter is simply energy in a polarised form, and that is what those scriptural words, "heaven and earth", are trying to describe. The first act of Creation is, and has to be, the creation of Universal Matter or Substance upon which the Word of God can act as a vibration, in order to build up form. Thus the second act of Creation is the coming of the Word, borne "on the wings of the dove" (the Holy Breath of Divine Love), into this universal and pure Matter (the Virgin Mary). The result of the vibration of the Word of God in the virgin Matter is the perfect Form or manifestation of God, which is Light, the Christos.

THE PIERCING AND RAISING OF THE SERPENT-DRAGON

For the furtherance of Creation - the evolution of self-consciousness or self-knowledge - the Light divides into three realms: the angelic, archetypal realm of the Christ Spirit; the human, self-conscious realm of the evolving Soul; and the elemental, experiential realm of the Body. Although there is a yet higher interpretation, nevertheless the serpent-dragon as used here represents the elemental matter of the Body realm - the polarised, locked-up energy of the ethers and lower elements. In the East it is called kundalini. This lifeforce has to be released from the Body realm and raised up to form the perfect soular life-form that perfectly images or "reflects" the spiritual Archetype of radiant Light. This is only done when rays of the spiritual Light, each an archetype that carries the pattern for a soular life-form, is brought down into elemental matter, stirring it up, releasing the kundalini energy and guiding that energy-matter or natural life-force to flow into an expression of the archetypal, angelic form. This is what the symbol of the sword piercing the dragon means: it is the descent of the Christ Spirit into elemental matter so that the lower self or "Body" may be transmuted, raised and built up into the beautiful soular form of light, the perfect human Soul.

MICHAEL-APOLLO AND BRITANNIA-ATHENA

So the serpent-dragon is first slain (from the viewpoint of the lower realm), then resurrected or raised up on the sword or spear of angelic light. The Archangel Michael is the usual angelic Archetype or "God" that is portrayed as wielding the sword or spear, but Britannia stands for the same - the feminine counterpart of Michael. The Greeks called these two Apollo and Pallas Athena respectively. Pallas Athena actually means "the striker, the shaker of the spear". That is what her name literally means, "the spear-shaker" or "Shake-spear". She, like Michael, shakes the spear and strikes it at the serpent of ignorance that writhes upon or in the ground of elemental matter, in order that that serpent may rise up her spear and become the wise serpent - the illumined soul. It is this great symbol that Francis Bacon took for his whole work, and the Shakespeare Plays simply portray this fact and the ways in which the spear of enlightenment works in the lower realm. The brotherhood of knight-Adepts knew (and still know) exactly what was meant and what it was all about, simply because of the use of that name in conjunction with other "signatures".

THE CHRIST DEGREES: THE GREATER MYSTERIES

Well, the Adept enters into the Higher Degrees, the real Greater Mysteries, the Christing Degrees or Degrees of Christhood, as he becomes a true Master or knight. The fifth initiation is the first Degree of Christhood, and the Adept learns to become a master of his total being. As such he controls the dragon energies of his transmuted lower self in order to use all the energy for the good of all. Whilst the soul is unfolding his mastership, he becomes more and more a healer in a great sense. You will find that healing is actually the training to become master of oneself, on all levels. In healing the soul learns how to use the energies that are pouring through him or her, both upwards from the earth and downwards in terms of the spiritual light or archetype. The conscious healer uses the blend of both these energies to give out healing. So the true knight is the healer and helper of mankind, on all levels, and he is the bard, the poet-musician *par excellence*. Western tradition calls such a Master, a "St. George", knight of the Rose and Golden Cross.

THE SOLAR BOAT (AIR-ETHER): UNIFICATION

When full mastery of all seven Rays or aspects of Light and life is achieved, the great Master enters into the sixth Degree of initiation, where is found an increasingly full revelation or illumination of the whole spectrum of Rays, blended as one blazing white Christ Light. The Master soul - the rose flame of the heart - opens fully out to embrace, comprehend and manifest the Christhood at all times. He becomes the true priest and seer - overseeing all evolution below him. The symbol for the sixth initiation is the solar boat. The great Master and priest of light is symbolically placed in a boat, the Ark or Argo of the knight-heroes, or navis (nave) of the Christian Church. He is cast out on the seas of Time, the seas of Eternity, the seas of Universal Matter, and he sails towards the Crystal Isle or Temple of the incarnate Spirit. The Crystal Isle represents the illumined Soul, the Soular Bride of the Spirit in which the Spirit is fully "married" and made manifest, the "New Jerusalem". In the ceremonies of old the initiate was actually put into a boat or coracle and shoved out onto a lake , river or sea, to drift for three days and nights. Eventually he would land on the beach of the holy and "royal" island. The initiate was not told where or how he was going to land, or that he would land. He had to do it in complete faith - a knowledge of and trust in God's protection and the good intention of His ministers (the Brotherhood giving him the experience of initiation). It could be quite an ordeal, especially when he was cast out to sea in an area of storms and treacherous currents.

ST. COLUMBA AND IONA

St. Columba went through this initiation when he went to Iona. He went through that Degree of initiation, at the particular level he was capable of, and the Druid priests put him into a coracle with his friends who were under-going the same Degree of initiation with him. They had to live together in that tiny coracle, tossed about on those rough seas off the west coast of Scotland, and land up on Iona, not knowing whether they were going to land or not. The hierophants of course had an intimate knowledge of the currents, and were able to use them for these purposes. Even now in Scotland there is a knowledge of the tides which will sweep a certain point on the mainland of Scotland and take a bottle or a boat out to sea at a certain time of the year and land it on St. Columba's Bay on Iona. But there are, of course, such things as storms and unforeseen changes in current. If such happenings occurred it was taken as being as the result of God's Will overriding the normal course of events. There was always an element of not knowing, and many initiates were perhaps drowned during the initiation ceremony. The Druid hierophants would always do their best to ensure a successful initiation ceremony, but always there was and must be the unknown Divine Providence and Fate. It is something that we must learn to accept, that we can never be master of everything; there being only one true Master, the Universal or Cosmic Christ, the Light of God.

What this solar boat symbolises, therefore, is that, when the initiate graduates from being a knight to become a priest, he has no fixed home as such, and must entirely "put his hand into the hand of God". The whole world, even the whole universe on a higher cycle, becomes his home. The priest-seer wanders around doing service wherever the Spirit calls, wherever he is needed. He has to be prepared to go anywhere, do anything, at any time, so he sails upon the Sea of Time.

THE ANKH, CROSS OF LIGHT AND LIFE (FIRE-ETHER): ENTHRONEMENT

The seventh initiation is that of the Sovereign, the Arch-Priest and Sun King, the Fully Illumined or Christed Soul. The great Master soul reaches the point of enthronement. Dionysus, or Jesus, ascends to heaven and sits on the "right hand" of the Father of Light. Then, as Lord of Compassion, he knows that there is only one thing left for him to do before any further evolution can take place for him. He knows that he must come down again to earth, to help redeem what is still left. And so, he returns again and again as Hermes, the Christed One, the fully Illumined One, the Messenger or Avatar of Light. Such a One comes in answer to world prayer, at special times of need and opportunity for evolution, to give more and yet more to life. But it should be realised that each time such a great One approaches close to our level of strife and inharmony, only a small part of his total Being can possibly be made manifest to us, sufficient for the purpose. The symbol of such a One who is all love, light and life, is the Ankh.

All these initiations operate on various levels and cycles of manifestation. In other words, we all in our lifetimes go through each of these degrees at a certain mundane level. We can think of every one of these seven Degrees as being repeated in a microcosmic way, giving seven lesser Degrees within each greater Degree. If we take white light, for instance, and put a prism in its path, that light can be split up into seven colours. If we take another prism and split up any one of those colours we find that we obtain seven more shades of colour out of that one colour; and so it goes on *ad infinitum*. So it is with the initiations. Each of these great initiations has in fact seven stages within it, which are equivalent to the same ladder of great initiations but on a lesser scale; and in each of those lesser seven Degrees there is another microcosmic seven, and so on - "wheels within wheels". We are all going through these initiations but at varying levels.

This is something that one needs to fully understand. The position of Druid, a "Holy One", for instance, represented the full Christhood or Illumination. Yet at the most but a very, very few ever reached a manifestation of the full Christhood (as far as it is possible in a single personality and physical body) and yet many individuals became Druids. This was because they were taking that seventh initiation on a lower level or cycle, and to a certain extent they had reached a sort of Christhood, but not in the full sense. Nevertheless a Druid had been through the experiences of the seven initiations in a complete cycle, and in future incarnations they will be able to do it on a slightly higher level still, and so on. Eventually such a soul and personality will reach the stage that the Master Jesus was (and is) at; and so will all of us. This hope was a promise that Jesus was able to give to us all, because it is part of the plan of God. One day we shall all eventually reach the stage of full Christhood; but we attain it step by step, and each step involves experiencing something of each of these initiations, in cycle after cycle.

Summary of the EIGHT SOLAR FESTIVALS

THE FESTIVAL OF PEACE

The Festival of Peace, called by the Celts Samh'in ("the Fire of Peace"), is concerned with the climax of the annual cycle of nature and of man's initiatory experience, and with the beginning of a new cycle. The fruit of one cycle is given to the earth in total sacrifice so that a new cycle may begin. Its seed enters the darkness of the ground in order to bring forth new and more abundant life after the Winter period is over. In the ancient British or Druidical calendar this was the start of the new year. In the Orphic mysteries this marked the Thesmophoria, the beginning of the Mystery cycle and preparation for initiation into the Lesser Mysteries of Purity; and, on a higher cycle, into the Greater Mysteries of Light.

THE FESTIVAL OF REBIRTH

The Festival of Rebirth, at Midwinter, celebrates the Winter Solstice, when the sun, which has been steadily waning or appearing lower in the skies each day since Midsummer, stops its apparent descent into eventual obscurity, pauses for three days and then begins to rise higher in the skies as each day proceeds towards the next Midsummer. The nights which had been growing longer at the expense of the daylight, threatening eventual and perpetual darkness with extinction of all life form, now begin to grow shorter and the days to lengthen, restoring light, warmth and vitality to the cold world. This Festival is the Christ Mass, when all God's love and light is sent again to the world to save it from the grave of death, to redeem and revivify nature, and to build that nature up into a fresh cycle of manifestation of the spirit. In the West we celebrate this Festival on the third day of the Solstice, when the days can first be seen to commence drawing out; and this is the day on which the child Horus is eternally reborn in the Egyptian Mysteries, and likewise the baby Jesus in the Hebraic-Christian Mysteries, and Iacchus-Dionysus in the Orphic Mysteries. The Festival of Rebirth is closely linked with the Festival of the Immaculate Conception, which precedes Christmas by a few weeks. The Festival of the Immaculate Conception of Our Lady is concerned with the preparation and purifying of the earth, and of our hearts and bodies, so that they provide a perfect or immaculate habitation in which the seed, both natural and spiritual, may properly germinate and begin the growth which manifests its hidden archetype or spirit. It is the germination which is associated with the Winter Solstice and Christmas, the birth (or rebirth) of the Christ child in the pure or virgin heart of a soul.

THE FESTIVAL OF DEDICATION

The Festival of Dedication, called by the Celts Imbolc and by the Christian Church Candlemas or Feast of Lights, is concerned in nature with the revealing of the sprouting seed as it pushes its head above the ground, and in man with the realisation of the flame of Christ light within the temple of his heart. At this stage the psyche as well as the virgin heart is now become sufficiently pure for the light within to be seen and recognised. With this realisation the soul dedicates itself to the service of that light, that it might eventually shine forth in full glory. Sometimes therefore the festival is called the Feast of the Purification of the Blessed Virgin Mary, or the Presentation of the Christ in the Temple. The festival marks the end of Winter and the beginning of Spring. In the Orphic mysteries the Lesser Eleusinia commenced at this point, the start of initiation.

THE FESTIVAL OF PROMISE

The Festival of Promise celebrates the Spring Equinox, and is intimately linked to the Passover and Paschal Feast. It is associated with Lady Day (the Annunciation of Our Lady), and is concerned with the vibrancy of reawakening life after the Winter sleep, the promise of love, and of a rich and fulfilled life ahead; and the excitement of the quest for this fulfillment surges up inside the seeker. A glimpse of future possibilities is seen, and a love-troth is made as the vision of the Bride or Virgin holding the Holy Grail is seen. The knight begins his quest in earnest; the youth his courting. The Western world celebrates the bestowal of this great vision or hope of redemption and fulfilment with the pageant of Easter.

THE FESTIVAL OF UNIFICATION

The Festival of Unification, called by the Celts Beltane ("the Fire of Bel" - "the Fire of Light, or Wisdom"), is intimately associated with the Wessak Festival, held at the time of the full moon when the sun rises in Taurus each year. The festival is concerned with the blessing and fertilisation of all nature with light so as to bring all to blossom and fruit. For man it marks his "Coming of Age" when he truly finds something of the Holy Grail, the ideal state of love, and celebrates the discovery in marriage and in consecration as priest-priestess. It is the original Whitsun or Pentecost, which reaches its climax in the Christ Festival of the following full moon. The emotions of man now being calm and pure, his physique sturdy, the Holy Spirit of Love-Wisdom is breathed out upon him to illumine his mind and to bring his heart to flower. The twin bonfires of opposite but complementary natures are lit, and man and woman walk between the flaming beacons to find their unity in the mystery of marriage. The festival marks the end of Spring and the beginning of Summer.

THE FESTIVAL OF JOY

The Festival of Joy, or Midsummer, celebrates the Summer Solstice, and is linked with the Christ Festival of the preceeding full moon (when the sun rises in Gemini). At the Christ Festival the Pentecost or great baptism of light is brought to a climax in a wonderful outpouring of divine Love-Wisdom, the Christ Love, upon the whole world. This ushers in the Midsummer Festival of Joy, when all nature rejoices in the light, showing forth its colourful and fragrant beauty, whilst man opens out his heart in love and joy. On this day Arthur, the Sun King and High Initiate of the Mysteries of Light, holds his Court, at which all true priests and priestesses, knights and maidens, gather in due order to rejoice in the light, transfigured by and bearing witness to the light. The single great bonfire of radiant illumination is lit, emblem of the perfect union and consummation of marriage, and Arthur's Court makes the circle of love around it, as one whole being, in the dance of joy and thanksgiving.

THE FESTIVAL OF TRANSFORMATION

The Festival of Transformation, called by the Celts Lugnasadh or Lammas (i.e. Hlaf-maesse, "Loaf Mass") marks the beginning of harvest, when the fields of corn are beginning to ripen and trees to bear fruit, and the "first-fruits" are taken and offered to God. This solar festival is linked with the full-moon festival when the sun rises in Leo, and for man it marks the culmination of his transfiguration by light, and his crowning with the Christ aura or corona of light, making of him a true initiate and priest-king. His soul endeavours have flowered and are bearing fruit, and he now contains the seed of immortal existence and the Christ glory, - his life being surrendered totally in service to guide and illumine the community that recognise and select him as their king, and to share their burdens. The festival marks the end of Summer and the beginning of Autumn.

THE FESTIVAL OF CONSUMMATION

The Festival of Consummation at the Autumnal Equinox is the ancient Michaelmas - the Feast of St. Michael and All Angels. It is also known as the Harvest Festival, when the ripe corn is reaped in the fields and left in sheaves to dry in the sun's rays. St. Michael is "He who is like unto God" - "the Countenance of the Lord". A soul who has become so beautiful and radiant that the light and love of God literally shines through his face, revealing the Secret Heart of God, is manifesting the Michaelic principle. Such a one is a true initiate, standing on the symbolic mount of crucifixion, a light to those around him, but knowing that he has to suffer - that his old, mature self has to die so that the real Christ self within can be born or released from its womb. He is the harvest - the plant that will shrivel and die in the fire of the Sun, and the seed that will be rescued and made into the Bread of Life. In the Orphic mysteries this marked the beginning of the Greater Eleusinia, which culminated at the Festival of Peace, the start of a new cycle - either of the Lesser Mysteries of the Virgin Mother (beginning with the Thesmophoria), or of the Greater Mysteries of the Son of Light.

THE FRANCIS BACON RESEARCH TRUST